The Model T Ford Car

Car

Its Construction, Parts, Operation and Repair –

A Mechanic's Illustrated Treatise on the Automobile from 1915

By Victor Wilfred Pagé

PANTIANOS
CLASSICS

Published by Pantianos Classics

ISBN-13: 978-1-78987-375-7

First published in 1915

Contents

Preface

There is only one make of motor vehicle in the world that is sold in large enough quantities to warrant the publication of a special treatise on its repair and maintenance and that is the Ford Model T. With the close of the 1915 season's business there will be at least 600,000 Ford cars of all types in use, perhaps more. Most of these cars have been sold to and are being operated by persons with but little mechanical knowledge and with no experience with the gas engine as an automobile power plant.

The maker's instruction book is excellent but it is necessarily brief as there is no opportunity for an extended exposition of principles involved. Many operators desire to know the first principles before studying the operation and repair. While considerable instruction is given in the writer's large work on motoring, "The Modern Gasoline Automobile," it is not possible to cover any specific make of car completely in a general treatise.

Many requests have been received from motorists for a book on the Ford car that would enter into elementary exposition more than the manufacturer's instruction book does and cover some of the points involved in repair and maintenance more completely.

The writer has operated a Ford car of his own for nearly four years and has had many other cars of the same make under observation. This has made the collection of much data pertaining to repair and maintenance from first hand sources possible and has resulted in an appreciation of some of the points about which more information could be imparted to advantage. Special photographs have been taken and drawings made to make the subject matter easy of comprehension, even to the student and non-mechanical owner. Needless to say, the general repairman and dealer will also find much of interest in this volume.

VICTOR W. PAGE.
October, 1915

Chapter One - The Ford Car, Its Parts and Their Functions

In order to have any subject easily understood by the layman, especially if it is a mechanical topic or one with which the public at large is not thoroughly familiar, it is always necessary to consider first of all the basic principles underlying the operation of the mechanism discussed. Those who are familiar with the subject to a degree may consider this matter superfluous because it is a review of things of which they already have some knowledge. The person who seeks information, especially the purchaser of an automobile who intends to operate it himself, in many cases does not have the slightest conception of mechanics. It is necessary to describe fully the various parts and how they operate and why they work as they do before any attempt is made to give suggestions for their care or operation.

In making repairs or looking for troubles the man who is familiar with the principles of action of the parts at fault is nearly always able to locate the trouble whereas those who are not posted on the methods of working are at a loss because they do not know where to start to look for derangements. The automobile has been the greatest popular mechanical educator ever devised, but it is a much simpler and less expensive process to acquire this knowledge by becoming familiar with the experience of others instead of learning all the points involved by the slow and uncertain process of actual personal experience. In preparing this treatise the writer believes that it will have more value for most of those it is desired to instruct if the assumption is made that the reader knows absolutely nothing regarding automobile construction. For this reason, the exposition starts with a description of the parts that are absolutely necessary to secure successful operation of any self-propelled road vehicle, then various units of the car discussed are described and their functions outlined. Endeavor has been made to present the information in a clear manner and to avoid technicalities. It is desirable, however, that mechanical terms be used and all parts called by their correct names, ready identification being provided by clear, lettered illustrations.

Important Components of Any Motor Car. — In this era of progress, one would hesitate to assert that the motor car had been perfected or it had reached a finality in design, though the experience of the last few years would justify one assuming that the principles of construction now applied so successfully may reasonably be considered permanent. The elements which have been proven essential to insure successful operation of all self-propelled conveyances may be easily defined as follows:

First: The endeavor of modern constructors is to make all operating parts of such material, size and strength, that the severe strains imposed by the rough nature of the average road surface will be resisted adequately and to secure endurance and serviceability under all possible conditions of operation.

Second: The mechanism should be as simple as it is possible to make it, as this promotes ease of repairing, facility in handling, and lessens the liability of trouble by reducing complication. The parts should be in proper proportion and arranged in such a manner relative to each other that one may be removed or replaced without disturbing other correlated appliances.

Third: The power furnished by the gasoline engine carried in the frame must be transmitted to the traction wheels or to the revolving shafts to which they are fastened with as little friction and power loss as is possible.

Fourth: The two driven wheels (preferably the rear ones) must be connected to some form of compensating or balance gear which enables each wheel to revolve independently of the other at times and at different velocities, because in turning corners the outer wheel describes a larger arc and consequently a longer path than the inner member. The differential gear was one of the most important elements which made for the successful development of the automobile.

Fifth: The steering should be done by the two front wheels which are carried at the ends of a yoke axle which is securely fastened to the chassis frame by means of the springs. The wheels are carried on steering knuckles which must be arranged to assume different angles when the vehicle is turning corners or deviates from a straight path in order to secure positive steering.

Sixth: Springs must be provided which will have sufficient strength and elasticity to neutralize vibration and allow for unevenness of the road surface by their yielding qualities and thus reduce body movement. In order to relieve the machinery, running gear and passengers of the inevitable vibration which obtains at even moderate speed on ordinary roads, the wheels should be provided with very resilient tires, preferably of the pneumatic or inflated forms for pleasure cars, and cushion or solid rubber on the heavier and slower-moving motor trucks.

Seventh: The gas supply to the motor, the ignition of the charge, and the continuation of the cycle of engine operations should be automatic and require no attention from the operator after the motor is once started. To secure continued operation, mechanical means must be provided for constant lubrication of all moving parts. All components which have movement relative to other parts should move with as little power loss by friction as possible, in order to conserve the available motor energy for tractive purposes. Anti-friction bearings of the ball or roller type should be employed on all rotating shafts in the power plant, transmission system, and in the wheels to save power.

Eighth: The center of gravity must be carried relatively low, which involves placing the body as close to the ground as practical considerations will permit. The wheel base, which is the distance between front and rear wheel centers, should be long, in order to secure the best result in tractive effort, steering and comfortable riding. The power plant and other essential mechanism should be carried on a frame which will be supported in such a manner that road shocks will not be transmitted to them and so coupled together that no frame distortion will produce disalignment of the driving shafts.

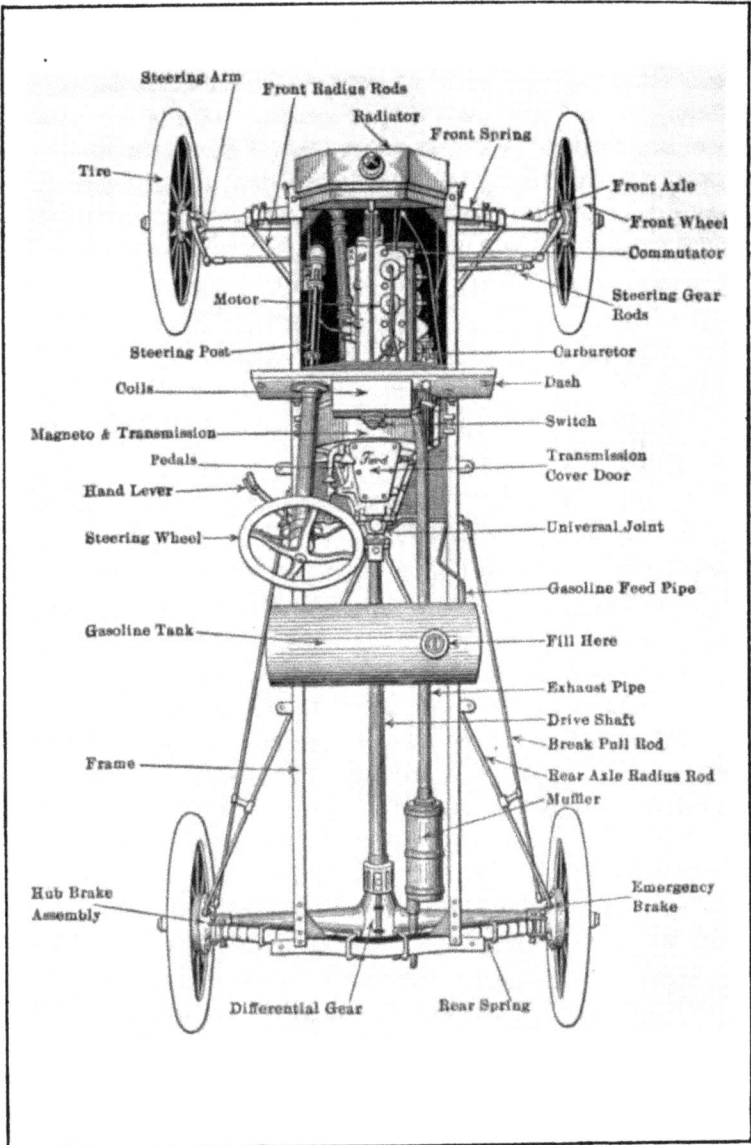

Fig. 1.—Plan View of the Ford Chassis Showing Relative Location of Important Components.

Ninth: The control elements must be designed with a view to easy handling. This means that the steering gear should be practically irreversible —

7

i.e., the hand wheel should not be affected by side movement of the front wheels, thus relieving the driver's arms of all undue strain while driving. Motor regulation should be by levers placed convenient to the driver's hands or feet, and gear, shifting should be accomplished without difficulty. Powerful brakes must be employed to insure positive check of vehicle motion whenever it is desired to bring the conveyance to a stop. It is evident that the levers through which the brakes are operated should be so proportioned that a minimum of effort on the part of the operator will serve to check the vehicle immediately.

Parts of Ford Automobile Chassis. — A brief explanation of the function of each part of the Ford gasoline car chassis depicted at Figs. 1 and 2 will serve to afford a better understanding of the construction of an automobile. The purpose of the front axle is not unlike that of a horse-drawn vehicle, but it is much different in construction. The wheels are installed on movable spindles, or steering knuckles, which are supported by yokes permitting one to move the wheels for steering rather than turning the entire axle on a fifth wheel, or jack-bolt arrangement, as in a horse-drawn vehicle. This axle is attached to the frame by spring member which allows a certain degree of movement without producing corresponding motion of the frame. The radiator, which is placed directly over the axle in front of the motor, is employed to hold the water used in keeping the engine cool, and is an important part of the heat-radiating system. The starting handle is a crank by which the motor crank shaft is given sufficient initial movement by the operator to carry the engine parts through one or more portions of the cycle of operations, this starting the engine. The tiebar joins the arms of the steering spindles on which the wheels revolve, and insures that these will swing together and in the same direction, either to the right or left. The steering link, often called the "drag link," connects one of the steering knuckles of the front axle with the steering gear.

The motor is a four-cylinder four-cycle type, to be described in proper sequence. The dash is a wooden partition placed back of the power plant to separate the engine space from the seating compartment. It is employed to support some of the auxiliary apparatus necessary to motor action or some of the control elements. The clutch is a device operated by a pedal, which permits the motor power to be coupled to the gearset, and from thence to the driving wheels, or interrupted at the will of the operator. It is used in starting and stopping the car, and whenever the speed is changed. The pedals are foot-operated levers, one of which releases the clutch and applies the slow speed; one is used to reverse the car, the other applies the running brake on the transmission. The motor control levers on the steering column are used in conjunction to vary the rotative speed of the motor, and thus regulate the energy produced in proportion to the work to be performed. The emergency brake lever applies a powerful braking effect when it is desired to stop the car quickly, and also when one wishes to lock the brakes if the car movement is arrested on a down grade. The steering wheel actuates the mechanism which moves the wheels to the right or left when one wishes to describe the circle, turn a comer, or otherwise deviate from a straight line.

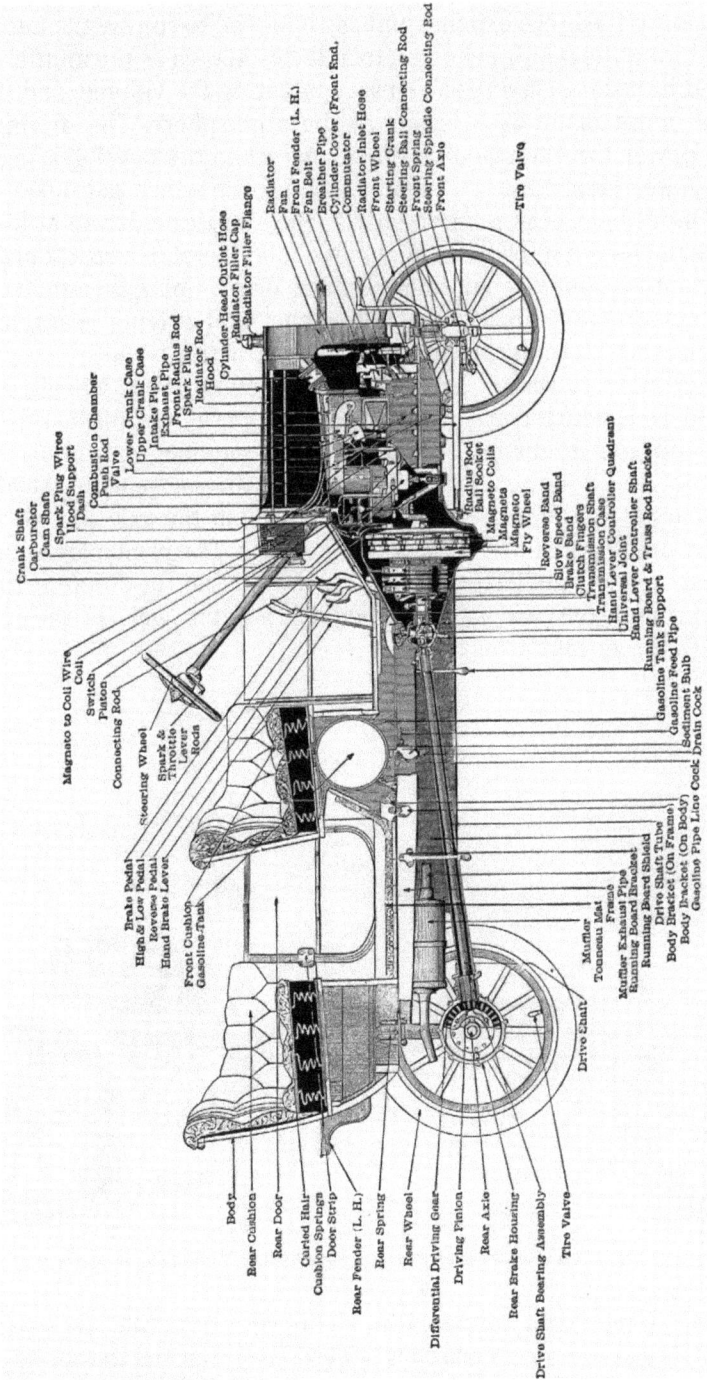

Fig. 2. - Sectional view of Ford Model T Touring Car Showing Construction of Chassis and Body Parts

The change speed gear is one of the most important elements of the power transmission system, and in connection with the clutch it is much used in operating and controlling the vehicle. The function of the frame has been previously described. The exhaust pipe is employed to convey the inert gases discharged from the motor cylinders to a device known as the muffler, which is designed to reduce gas pressure by augmenting the volume, and thus diminish the noise made as it issues to the atmosphere. The driving shaft transmits power from the change speed gearset to the bevel gearing in the rear axle. A universal joint is a positive connection which permits a certain degree of movement between two shafts which must be driven at the same speed. One or the other, or both, may move in a lateral or vertical plane to a limited extent without interrupting the drive or cramping the moving parts. The rear construction houses the differential and driving gears, and the shafts or axles which transmit the power to the traction wheels.

Brakes are used to retard, or stop the movement of the wheels, and are operated by rods which transmit the force the operator applies at the brake pedal or hand lever to the brake bands. Torque members or radius rods are used to maintain a definite relation between the driving gears in the axle and those in gearset, and to take the driving thrusts off the axle and the strains imposed by braking and driving from the springs. The principles underlying operation of each of the parts shown and the number of different forms in which they may exist, will be described more extensively in the chapters dealing with the various groups.

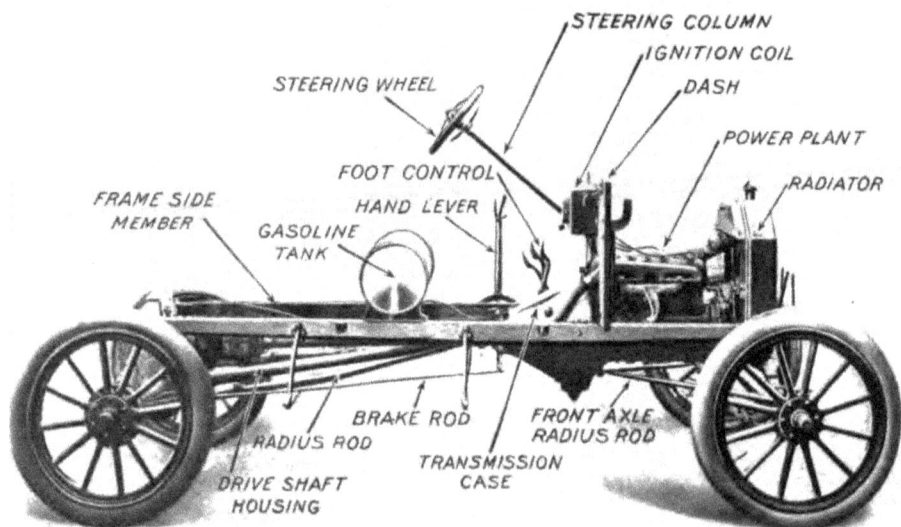

Fig. 3.—Side View of Stripped Ford Chassis Showing Valve Side of Motor.

The Ford Three Point Suspension. — In order to permit the sale of cars of good quality at moderate prices it is necessary that the design be simplified to a point where assembly cost during processes incidental to manufacture will be kept at a minimum. This elimination of unnecessary parts and the endeavor to simplify the assembly in order to reduce manufacturing expense really acts in favor of the purchaser because it is easier to maintain, operate and repair "a simple car than one having a greater number of parts and more complicated mechanism. In the design of the Ford chassis the main points attained have been simplicity and lightness without sacrifice of strength or endurance. The Ford car may be said to consist of four really essential groups, each of which is supported or joined to the other members by a three point suspension system. These component parts are the front axle group, the power plant assembly, the rear axle group and the frame.

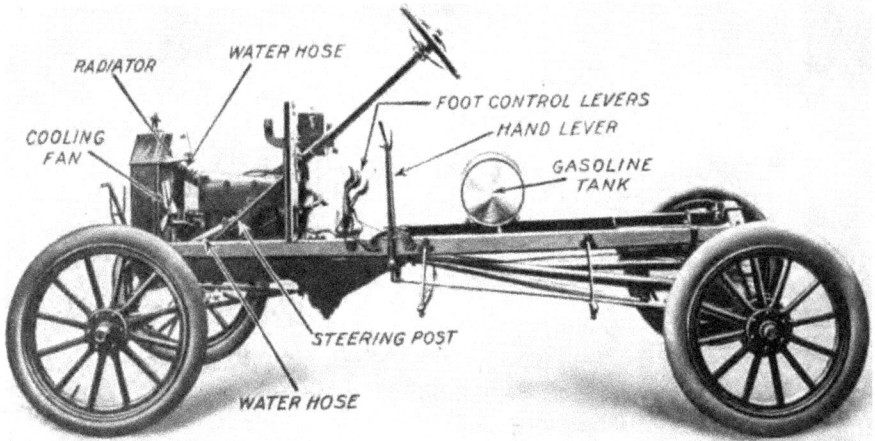

Fig. 4.—Control Side of Stripped Ford Model T Chassis.

The method of supporting the power plant by three points is clearly shown at Fig. 5, as is also the system of attaching the rear axle to the frame. A direct front view of the chassis at Fig. 6 shows the front axle construction and the method of supporting the front end of the frame. A direct rear view of the stripped chassis shown at Fig. 7 outlines the method of supporting the frame by a single cross spring of peculiar form and taken in conjunction with the plan view at Fig. 5, shows clearly the method of installing the rear axle to obtain a three point suspension of this member as well. There is a sound engineering reason for the three point system which is now widely followed in many automobiles. In the first automobiles, and in fact in many of the cars built to-day, the four point suspension system is followed in supporting the power plant to the frame. With the four point construction the power plant is provided with four supporting arms usually cast integrally with the engine base, two of these being on each side of the crank case. They are usually of such length that they can be bolted to the frame side member or rest on suitable supporting brackets riveted to the frame side or carried by the subframe. This makes a very rigid construction when the bolts are tightened

11

down and the engine bed firmly secured to the frame. While this method of support is very strong, it has the disadvantage of resisting any tendency of the frame twisting when the car is operated on unfavorable highway surfaces or when some one of the wheels passes over an obstruction or drops into a hollow in the road. These twisting strains stress the crank case arms arid very often will break them off.

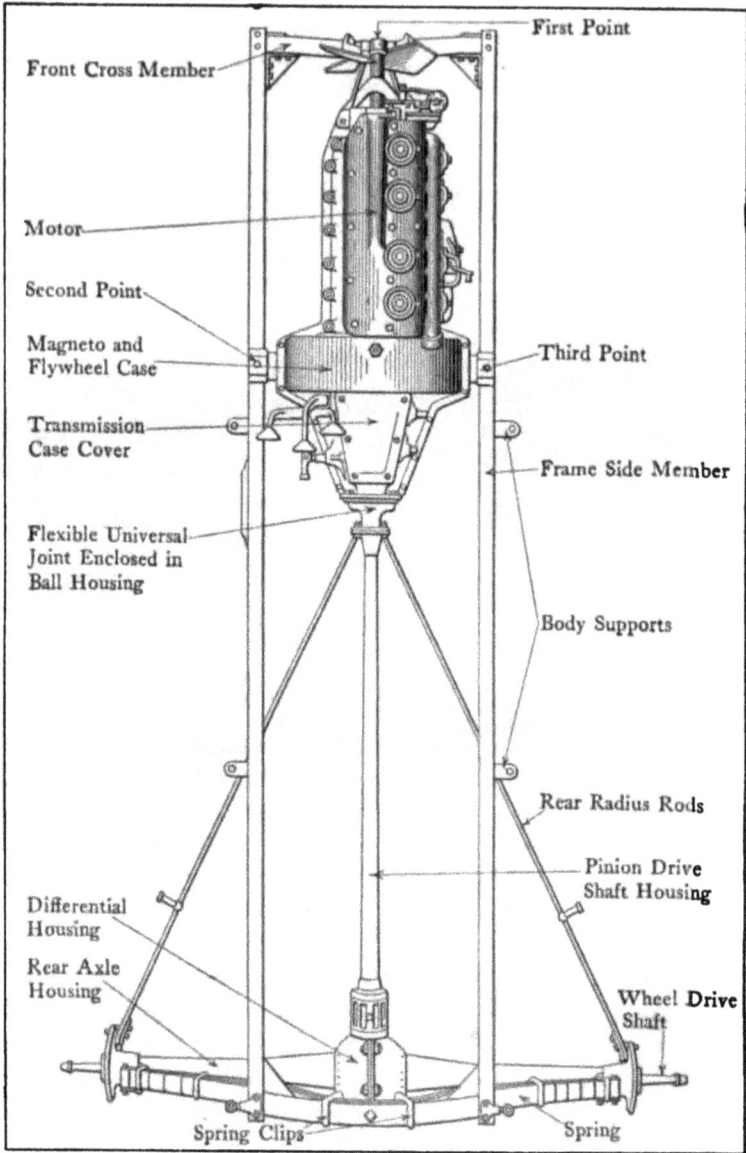

Fig. 5.—Plan View of Ford Frame with Power Plant and Rear Axle in Place Showing Three Point Suspension Principle Utilized in This Design.

It was to eliminate this that the three point suspension was invented. The four point suspension is found in most cases in cars where the engine is a separate member from the change speed gearing. Where the change speed mechanism is incorporated as a portion of the power plant, as is true of the Ford construction, it is possible to suspend the engine on: three points, as shown at Fig. 5. The first point is at the front end of the motor and consists of a turned cylindrical bearing resting in a trunnion block carried by the front cross member. The second and third points are provided by supporting brackets or arms of pressed steel which are securely riveted and brazed to each side of the flywheel housing. These arms rest upon the frame side member and are rigidly secured to the frame rail by bolts and lock. nuts. Wooden blocks are placed between the frame side channels and one bolt passes through the wood block horizontally, while another passes through the top and bottom of the frame side member vertically.

There is some difference in construction between the two points of support attached to the flywheel housing and the single or first point of support at the front end of the engine. While this holds the motor firmly in place, it permits a certain rocking action or movement of either front corner of the frame without likewise affecting the power plant. The front end bearing rests in a trunnion box made with a cap much in the same manner as one of the main crank shaft bearings of the engine. This is made in two sections, the lower or bed section being bolted to the front cross member of the frame, while the upper half, which is removable when it is desired to take out the power plant, is called the crank case front bearing cap, and is bolted to the lower member. By having a trunnion joint the twist imparted to the frame by inequalities of the road surface are not conveyed to the crank case as with a four point extension. For example, if the right wheel is raised six inches and the left wheel drops to the same amount, there will be a difference in level of one foot between them. This condition often obtains when driving a car on rutty roads, and' in this case the front cross member of the frame moves on the crank case bearing, but does not twist the motor to any extent from its normal horizontal position.

The front axle is also supported at three points, two of these, which are shown in Fig. 6, being to the front spring by means of shackle links attached to suitable forged hanger members bolted to the axle. This makes two points of support for the front spring. As the cross spring is not of such construction that it will push the front axle as do the semi-elliptic springs used in cars of conventional design which are placed on each side of the frame running parallel to the frame side member, and having the axle fastened at their centers; it is necessary to provide a radius or push rod construction which is in the form of a V-shaped member of steel tubing terminating in a ball at its apex and attached to the front spring supporting members at its ends. This triangular radius member serves to take the push of the chassis and move the axle forward as the car is driven in that direction. The ball rests in a socket member attached to the lower part of the flywheel casing of the engine as clearly shown at Fig. 3. It will be apparent that this method of support permits the front axle to move up or down or for one end to be higher than the other without tending to twist the frame as much as would be the case if the usual system of semi-elliptic front springs was used. A certain amount of twisting is unavoidable so the front axle three point suspension in connection with that of the motor insures that no strains will come on the crank case because of this varying frame distortion.

The rear axle assembly which also includes the driving shaft and its supporting housing is also fastened to the frame by three points. Two of these are at the axle where the single rear spring member is shackled to drop-forged steel hangers secured to brake shoe retaining plates while the third point of support is at a ball joint member attached to the rear of the transmission case. This ball joint serves to enclose the universal joint and performs the same function for the rear end of the car that the ball joint on the front end of the flywheel case performs for the front end of the car. The rear radius rod system is also triangular in shape, the rod members extending from the brake carrying castings on the ends of the axles to the flange fitting just back of the casting forming the ball part of the joint at the front end of the pinion drive shaft housing. This construction is clearly outlined in sectional view of the complete car shown on folding plate Fig. 2, and also in the plan view, Fig. 5. It will be apparent that this three point support of the front axle, the power plant and the rear axle makes for a construction of great flexibility which combines the very desirable element of strength and endurance without the undesirable one of excessive weight. The lightness and ability of the frame and chassis mechanism to endure under the twists and strains incidental to rough road work has made the Ford car a practical conveyance for unfavorable, as well as favorable, road conditions, and enhances the comfort of the occupants under all conditions. As an assembling proposition the three point suspension system is one that lends itself readily to quantity production, and if it is easily put together it is equally true that taking it apart will offer no difficulty. For example, if it is desired to remove tie rear axle or the front axle from the Ford chassis the 1st step is to undo the ball and socket joint at the end of the radius rod and to remove one half of each spring hackle in the case of the front axle, this setting it entirely free and permitting its easy withdrawal. In addition to undoing the front hall housing at the end of the pinion rive shaft housing of the rear axle and the spring shackle lember it is also necessary to uncouple the brake rods operating the rear wheel brakes at the point where they attach to the cross shaft actuated by the hand lever (early shown in Fig. 1.

Frame Assembly Details. — The frame of the Ford car is a very simple member, and owing to the three point system of attachment of components is made light and flexible to a degree. It consists essentially of two long straight side members and front and rear cross members. The material used for the side members is a channel section pressed steel, the frame side rails being about 100 inches long. The cross members are not straight, as is the case of the side members, but are bent as indicated Figs. 6 and 7. The front cross member is bent down to offer a support for the simple semi-elliptic cross spring while the rear member has an upward bend to fit the arch of the rear spring. The corners of the frame are securely braced by gusset plates or re-enforcing brackets, as indicted in Fig. 5. These are securely riveted to the frame side and cross members, the cross members also being attached to the side rails by hot riveting. Body supporting brackets are attached to the side rails to which the body portion is fastened by means of bolts. These are held in place by riveting, as are the running board supporting irons.

Fig. 7.—Rear View of Ford Model T Chassis Depicting Method of Attaching Rear Construction to Frame.

Spring Construction. — The first point that strikes the person examining the Ford car carefully for the first time, and one which serves as a ready means of identification, is the method of springing employed which is unconventional, though exceedingly practical. The springs of practically all automobiles are of the laminated type and are a built-up construction composed of long, flat sections of tempered high carbon or alloy steel that are called "leaves." The material used in the Ford springs is vanadium steel, an alloy which is said to possess greater endurance to continual deflection and rebound than ordinary high carbon steel. The Ford front spring is a semielliptic member, which term is applied to the spring according to the arc of an ellipse in which they are formed. A full elliptic spring would be one composed of two semielliptic members placed in such relation that they would produce a flattened circle or ellipse. A spring which is composed of one-half of a full elliptic member is called a semi-elliptic spring.

In common practice the center portion of the semielliptic spring rests on the axle of the car, the front end or eye being secured to the spring horn by a bolt while the rear end is attached to the frame by a shackle or hinge joint which is necessary to provide for the lengthening of the spring as it deflects under load, the tendency of the load being to straighten out the arch. In the Ford car the eyes of the front spring are shackled to the hangers on the front axle while the center portion is securely held inside of the channel section of the front frame cross member by means of U-shaped spring clips. This means that the semi-elliptic spring is inverted from the position it usually occupies on other cars. The front springs consist of seven leaves or laminations of graduated length, the longest being at the bottom and having eye members formed at each end, the shortest leaf being placed at the top. The seven leaves are held together by a tie bolt passing through their centers, which keeps the spring in shape whenever it is desired to unloosen the spring clips to -remove it from the frame. Rebound clips are also provided to keep the spring leaves from spreading apart under violent upward throw of the chassis frame. In order not to restrict free movement of the spring both front and rear shackles are provided with oil cups so that the shackle holts may be kept properly lubricated at all times. It will be apparent, however, that spring movement must necessarily be limited, and that the more resilient a spring is the greater the degree of movement. The springs are constantly in action when the vehicle is used, and practically all of the comfort of the occupants is dependent upon them.

The rear spring shown at Fig. 7 is also a cross member, but it is not of the same semi-elliptic form the front spring is. It has a decided arch at its center, designed to conform to the arched portion of the rear cross member, to which the center of the rear spring is secured by substantial spring clips. The spring perches or hangers are drop forgings similar in form to the front members, and have a boss pierced with a hole through which the bolt of the spring hanger or shackle member passes. The Ford rear spring perches are securely held in the brake retaining plate castings at the ends of the axle.

The Ford Body. — Practically the only part of an automobile that resembles in any way the horse-drawn vehicles they supercede is the body or carriage work portion employed for conveying the passengers. The body of the Ford car is clearly shown at Fig. 2. The sectional view gives an idea of the tufted upholstery and the method of constructing the seat cushions. The body is usually composed of a framework of wood, to which formed steel sheets are fastened, these being termed panels. By the development of large stamping presses it is possible to form the entire panel of a rear seat member, for instance, out of a single sheet of steel so that practically no fitting is necessary except to fasten it securely to the frame by means of small screws or nails. The wooden frame members serve as supports for the" floor boards, and also for the seat cushions. The doors by which the front or rear compartments are reached are light wooden framework covered with sheet metal and secured to the main body frame by simple hinges. The upholstering of the Ford car is a fabric made in imitation of leather, which is tufted and filled with curled hair in the fashion of regular carriage upholstery.

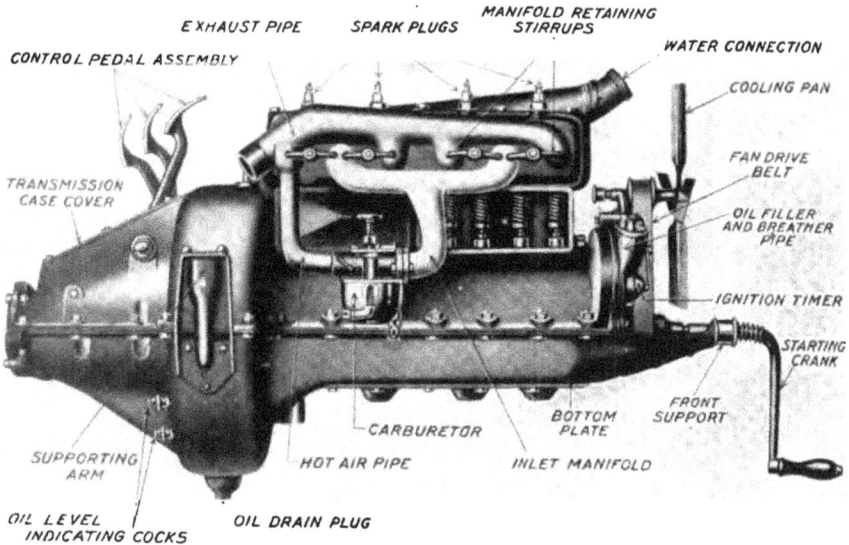

Fig. 8.—Valve Side of the Ford Model T Unit Power Plant Showing Manifolds, Carburetor and Interior of One of the Valve Spring Chambers.

The seat cushions are provided with a large number of coiled springs which are intended to support the passenger's weight and to supplement the action of the harsher-acting, stiffer chassis-supporting springs. These spiral springs absorb many of the minor shocks which would interfere with the comfort of the passengers if no springs were provided. It will be apparent that the passenger has three independent resilient supporting members between his body and the road. The pneumatic tires on the wheels form the primary shock-absorbing members, while the chassis springs are the secondary shock-absorbing members. The seat cushion springs and the curled hair padding are the final check against road vibration. In the Ford body the front floor boards are entirely removable, if desired, in order to gain access to the change speed gearing and the universal joint casing which are immediately under them. When the front seat cushion is lifted out, a hinged door permits of reaching the gasoline tank for filling. Lifting the rear cushion discloses another hinged door which provides communication to a compartment extending the full width of the rear seat suitable for carrying tools,

17

supplies and various articles of equipment. The windshield and top are not shown in illustration, though these are fastened to the body member. The sills of the body are provided with brackets of pressed steel, which are securely bolted to the wood and which rest on the body supporting brackets riveted to the car frame. The front end of the body is attached to the dashboard by angle irons. When it is desired to give the chassis a thorough overhauling it is not difficult to remove the body by loosening all of the body-retaining bolts, also the members holding the body to the dashboard assembly, which remains on the frame after the body is taken off.

The Ford Power Plant. — The motive power of the Ford car consist of what is known as a "unit power plant," incorporating the engine or power-producing member and the change speed gearing and clutch or power transmitting member. A side view of the engine is shown at Fig. 8, this showing clearly the parts that are readily discernible from the outside. The engine is a four-cylinder type, having the cylinders cast in one block which is integral with the top half of the engine crank case. As all the valves are on one side of the cylinder, it is known as an "L block" construction. The bore of the cylinder is 3¾", the stroke of the piston is 4" The engine is capable of attaining high rotative speed and will develop well over 20 h. p.

The advantage of including the change speed gearing and clutch in a unit with the engine is that there is no danger of loss of alignment of these members due to frame distortion as is possible when the power-producing and transmission elements are separate units, each having its independent means of support. The parts of the Ford power plant assembly are lined up properly when the engine is built, and there is no possibility of loss of this alignment until the engine has been in service a long enough time so that the bearings wear and permit the parts to get out of line. Long before this lack of alignment becomes serious the motorist will be warned that the bearings require refitting by noisy action of the power plant or change speed gearing.

The power plant, as shown, while complete in itself would not be operative unless a number of auxiliary mechanisms and devices are provided. In order to insure engine action it is necessary to supply the cylinders with a combustible gas, which is done by the carburetion system. The gas must be exploded in the cylinder to produce power, which is the function of the ignition group. In order to keep the engine at the proper working temperature and reduce internal friction, lubrication means must be provided, while a water-cooling system prevents the cylinders and combustion head from overheating. The basic principles of engine operation, also those underlying the action of the auxiliary groups, will be fully described in proper sequence.

Chapter Two - The Engine and Auxiliary Groups

The principles of action of all internal combustion engines are easily understood if one compares the effect produced by the explosion of the gas in the interior of the engine with the known effect obtained by firing any other explosive, such as gunpowder. Gasoline when mixed with air and compressed is highly explosive, and can be easily ignited by an electric spark. An explosion results from violent expansion which occurs where confined gases are fired. Combustion may exist in a number of different states, slow combustion, as the rotting of wood or rusting of iron; fast combustion, such as the burning of wood, coal or other fuels, and instantaneous combustion, which is that produced by igniting gunpowder or other explosives.

The manner in which a compressed gas can be made to give power may be easily understood by using as an illustration the difference between the noise of a firecracker that is in good condition and the "fizz" which results when an unexploded fire cracker is cracked open by an economical child and the gunpowder thus exposed is ignited. The reason we have noise when the fire cracker explodes is because the rapid combustion of the gunpowder in the confined spaces of the firecracker interior bursts the containing walls and produces a loud noise. In other words, the gunpowder has been compacted or compressed to a degree before ignition. When the powder is lighted in the firecracker that has been broken open, there is no noise to speak of because there is nothing to restrain the gases produced by burning the powder.

Another illustration that simplifies the method of operation of a gas engine is to compare it with a muzzle-loading gun or cannon. After the charge of gunpowder is introduced at the open end of the barrel it is necessary to ram this tightly in place in order to compact it before the shot is introduced. After the fuse is lighted or firing pin depressed the powder explodes with great energy and as the parts of the gun barrel surrounding the combustion chamber are sufficiently strong to withstand the force of the explosion, the movable member or shot is violently ejected from the gun barrel. If the same amount of gunpowder used to charge the gun was placed on a piece of paper and set afire, the only result would be a sudden flash or flame that would not be accompanied by an explosion, nor would it have any appreciable force because the gases are not confined.

In any internal combustion engine the gas charge is first drawn into a cylinder, which may be compared with the barrel of a gun by the suction effect of a downwardly moving piston member, and when the piston has reached the limit of its travel in one direction its motion is reversed and it moves back and starts to compress the gas previously inspired. As soon as the gas is properly compressed, which means that its volume has been reduced and its pressure increased, an electric spark takes place in the interior of the combustion chamber and the resulting explosion of the gas sends the piston violently downward, and this motion, through the medium of a connecting rod, imparts a rotary movement to a crank member.

The difference between an internal combustion engine and an external combustion engine is easily understood if one knows that the steam engine operates on the latter principle. With the steam engine, power is derived by admitting steam to the cylinder, which is obtained from a separate boiler member. This is produced by the evaporation of water by a fire under the boiler. It is evident that the power is really obtained by the combustion of fuel, the steam being only a flexible medium that has transformed the heat of the fire into the power. The steam bears against a movable piston member in the steam engine cylinder, and the movement of the piston is transformed into mechanical energy at the crank shaft in the same manner as in a gas engine. Wherever the fuel is burnt directly in the cylinder instead of under another part of the power plant so that its heat may be expended directly to produce piston movement with minimum power loss, the engine is called an "internal combustion engine."

How All Automobile Engines Work. — It is evident that burning powder in the air will produce a certain amount of energy, but as the explosion takes place in the open there will be nothing to restrain the pressure, and just as soon as the powder is lighted, any energy evolved by the combustion is dissipated into the atmosphere instead of the force being directed against a yielding member such as a shot. This bullet is forced out of the gun barrel, not only by the gas pressure which results as soon as the powder is exploded, but also by the expansion of the gases generated by combustion which tends to accelerate its motion toward the open end of the barrel. As the shot moves toward the end and the gas occupies more space its pressure becomes less, and when the ball leaves the mouth of the gun there is very little power remaining in the moving gas. There is sufficient pressure, however, so that the gas rushes out of the interior and the barrel is thus cleared of inert products which have no more useful force. The action of a modem repeating rifle is somewhat different than that of a muzzle-loader, because the powder is already compressed in metal shells which are introduced at the breech of the gun instead of at the muzzle. A number of shells are carried in a magazine, and after one of these explodes the recoil due to the explosion of the gas supplies another charged shell to the breech and the operation of firing the gun may be repeated as long as the supply of ammunition in the magazine lasts.

The modern four-cycle gasoline engine follows the action of both the old type muzzle-loader and the more modern form in which the shell is introduced at the breech. Referring to sketches at Fig. 9, we can compare the action of a simple four-stroke engine with that of a gun. The principal elements of a gas engine shown are not difficult to identify, and their functions are easily defined. In place of the barrel of the gun one has a smoothly machined cylinder in which a movable barrel-shaped element fitting the bore closely

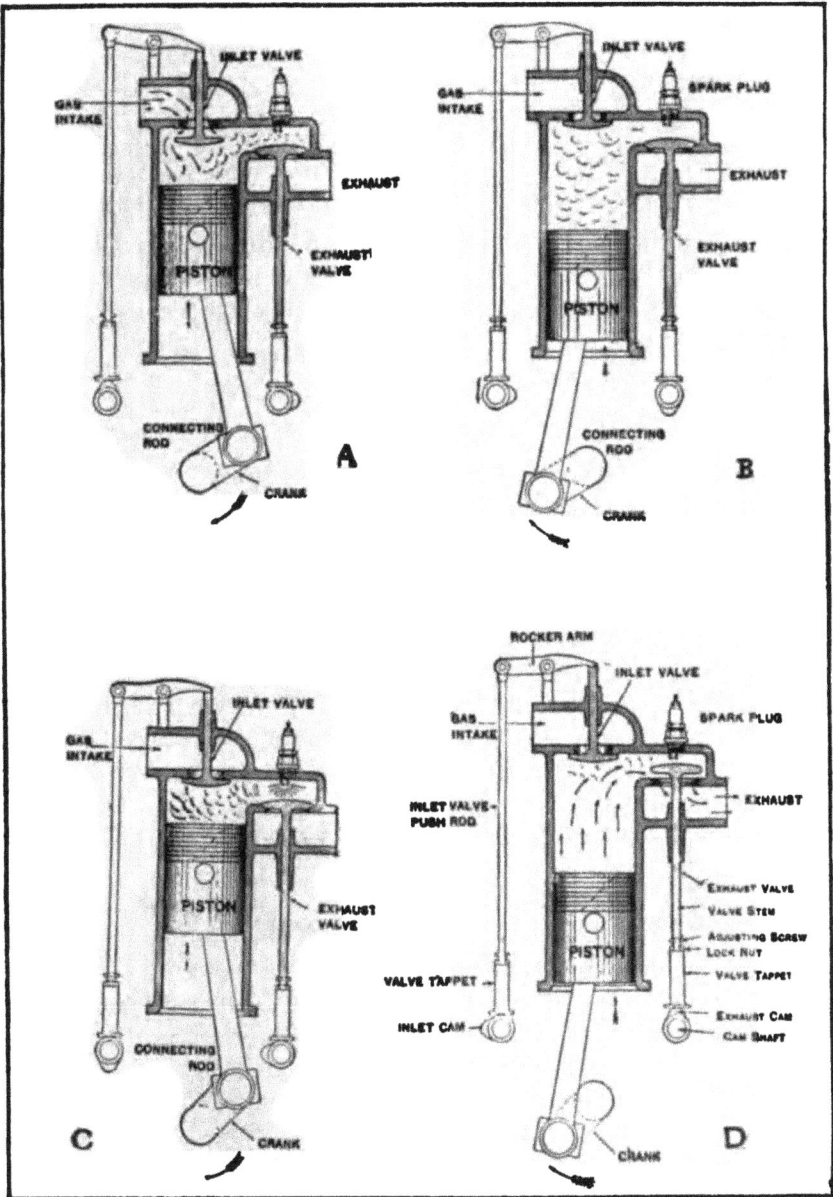

Fig. 9.—Simple Diagrams Showing the Various Cycles of Operation Necessary to Obtain an Explosion in the Four Stroke Gasoline Engine Cylinder. A—Suction. B—Compression. C—Expansion. D—Exhaust.

may be likened to a bullet or shot. It differs in an important respect, however, as while the shot is discharged from the mouth of the gun the piston member sliding inside of the cylinder cannot leave it, as its movements back and forth from the open to the closed end, and back again, are limited by simple mechanical connection or linkage which comprises a crank and connecting rod. It is by this means that the reciprocating movement of the piston is transformed into a rotary motion of the crank shaft. The flywheel is a heavy member attached to the crank shaft which has energy stored in its rim as it revolves, and the momentum of this revolving mass tends to equalize the intermittent pushes on the piston head produced by the succeeding explosions of gasoline vapor in the cylinder. If any explosive is placed in the chamber formed between the piston and closed end of the cylinder and exploded, the piston would be the only part that would yield to the pressure which would produce a downward movement. As the piston is forced down, the crank shaft is turned by the connecting rod. This part is hinged at both ends, so it is free to oscillate as the crank turns, and thus the piston may slide back and forth, while the crank shaft is rotating or describing a curvilinear path.

In addition to the simple elements described it is evident that a gasoline engine must have other parts. The most important of these are the valves, of which there are two to each cylinder. One closes the passage connecting to the gas supply and opens during one stroke of the piston in order to let the explosive gas into the combustion chamber. The other member, or exhaust valve, serves as a cover for the opening through which the burnt gases can leave the cylinder after their work is done. The spark plug is a simple device which may be compared to the percussion cap of a gun. It permits one to produce an electric spark in the cylinder when the piston is at the best point to utilize the pressure which obtains when the compressed gas is fired. The valves are open one at a time, the inlet valve being depressed from its seat while the cylinder is filling and the exhaust valve is opened when the cylinder is being cleared. They are normally kept seated by means of compression springs. In the simple motor shown at Fig. 9, the inlet valve is operated by means of a pivoted rocker arm moved by a cam which turns at half the speed of the crank shaft. The exhaust valve operates in a similar manner, as will be explained in proper sequence.

Considering the view shown at Fig. 9, A, the first necessary operation is charging the cylinder with explosive material. The piston is at the top of its stroke and it moves toward the open end of the cylinder. The engine works as a pump and the piston draws in a charge of combustible gas through the open intake valve which is in connection with the vaporizer or device which furnishes the gas. The valve opening is assisted by a light vacuiun or suction existing when the piston has traveled down a certain portion of its stroke, this supplementing the cam action as the outside air pressure is greater than that in the cylinder. The mechanical pressure produced by the cam is greater than the tension of the spring, which tends to keep the valve closed and that member is depressed from its seat and gas drawn in by the piston. At the end of the intake stroke, the start of which is shown at Fig. 9, A, and after the cylinder has filled with gas, the pressure inside and outside is the same, the cam pressure is released, and the valve spring closes the intake valve.

As the exhaust valve spring is very strong, this member has not been lifted from its seat by the difference in pressure during the suction stroke. The exhaust valve is opened by mechanical means solely and only when operated by the cam and push-rod mechanism. The condition in the cylinder of the gas engine after the piston has reached the bottom of its stroke is very much the same that which obtains in a gun of the muzzle-loading type after the explosive charge has been introduced. We have learned that to obtain power from gunpowder that it was necessary to compact it firmly in the combustion chamber of the gun. The gasoline gas which has been taken into the engine cylinder must also be compressed before it is ignited in order to obtain power. It is compacted into one-third or one-quarter of its former volume, and whereas its pressure is about fifteen pounds per square inch before the volume is reduced, at the end of the compression stroke of the piston the pressure will be increased to forty-five, sixty, and even seventy-five pounds per square inch. At the end of this compression stroke, the start of which is shown at B, Fig. 9, the conditions in the engine cylinder are the same as those which prevail in the barrel of the gun after the powder has been tightly rammed in the closed end of the gun barrel and the wadding and ball forced in on top of it. The explosion of the gas by the electric spark is shown at C, while the beginning of the exhaust stroke is depicted at D.

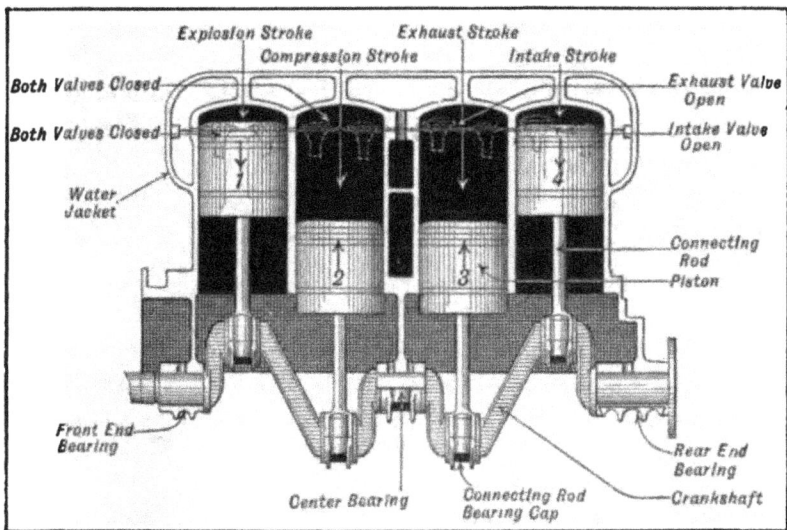

Fig. 10.—Diagram Showing the Relation of the Pistons and Crank Shaft Throws of the Ford Four Cylinder Motor when Piston No. 1 is about to Receive the Force of Gas Exploded in the Combustion Chamber.

With a one-cylinder, four-cycle engine it will be apparent that we have one useful power stroke in every four strokes of the piston. These four reciprocating movements are transformed into two complete rotary movements of the crank shaft to which the connecting rod is fastened. If we have a two-cylinder engine we can obtain one explosion for each revolution of the fly wheel or crank shaft and a more even turning effort and steadier power application will result. By using four cylinders it is possible to have one delivering power right after its neighbor leaves off and so a reasonably constant turning moment or more steady power delivery is possible than can be obtained with either a one or two-cylinder engine. In the Ford four-cylinder engine one obtains two explosions for each revolution of the flywheel. The position of the various pistons and their relation to each other when the front cylinder is about to fire is clearly shown at Fig. 10. Pistons Nos. 1 and 4 are at the top of the stroke as the crank pins operating them are on the same plane. Pistons Nos. 2 and 3 are at the bottom of the stroke as their crank pins are on the same line. It will he evident that the crank shaft of a four-cylinder engine must have four crank throws, one for each piston member.

In the Ford engine the firing order is 1, 2, 4, 3, which means that the front cylinder fires first, then the second cylinder, then the fourth cylinder, and lastly the third cylinder. The four operations of the cycle — suction, compression, explosion and exhaust — are repeated in regular order in each of the four cylinders according to the firing order. The pistons move downward during the explosion and intake strokes and they move upward during the compression and exhaust strokes. As can be readily ascertained from Fig. 10, when piston No. 1 is at the top of its stroke with the gas above it in the combustion chamber fully compressed and ready for ignition, piston No. 2 is at the bottom of the suction stroke and just at the beginning of its compression stroke. Both valves in cylinders Nos. 1 and 2 are closed. When the piston in cylinder No. 3 starts moving up, it will force the burnt gas out of cylinder No. 3, through the open exhaust valve. Piston No. 3 is starting on its exhaust stroke while piston No. 2 is moving in the same direction, only performing its compression stroke. Piston No. 4 is moving in the same direction as piston No. 1, which is about to be forced down by the pressure of the exploded gas, and as the inlet valve is opened in cylinder No. 4 a charge of gas will be drawn in from the carburettor when the piston moves down.

We have different conditions in the various cylinders, as follows: During the first half of the first revolution of the motor crank shaft, piston No. 1 is on its explosion stroke. No. 2 is compressing the gas, piston No. 3 is pushing out burned gas, while piston No. 4 is drawing in a fresh charge. During the second half of the first revolution of the crank shaft, during which the pistons reverse from the position shown in Fig. 10, piston No. 1 is about to clear the cylinder of burnt gas, piston No. 2 is to receive the force of the explosion, piston No. 3 is to draw in a fresh charge, while piston No. 4 is about to compress the gas taken in on the preceding down stroke. At the beginning of the first half of the second revolution of the crank shaft the pistons are again in the positions shown at Fig. 10, though the functions they are performing are not the same. Piston No. 1 is about to draw in a fresh charge of gas. Piston No. 2 is about to force out exhaust gas. Piston No. 3 is about to compress a charge, while piston No. 4 is in position to be acted on by the explosion that is to take place in cylinder No. 4 as soon as the gas is ignited. During the second half of the second revolution, piston No. 1 is compressing a charge, piston No. 2 is sucking in gas, piston No. 3 is being acted on by an explosion, while cylinder No. 4, in which a charge has exploded on the previous stroke of the piston, is being cleared of burnt gas. It will be apparent that by using four cylinders, as in the Ford motor, practically steady power application is obtained.

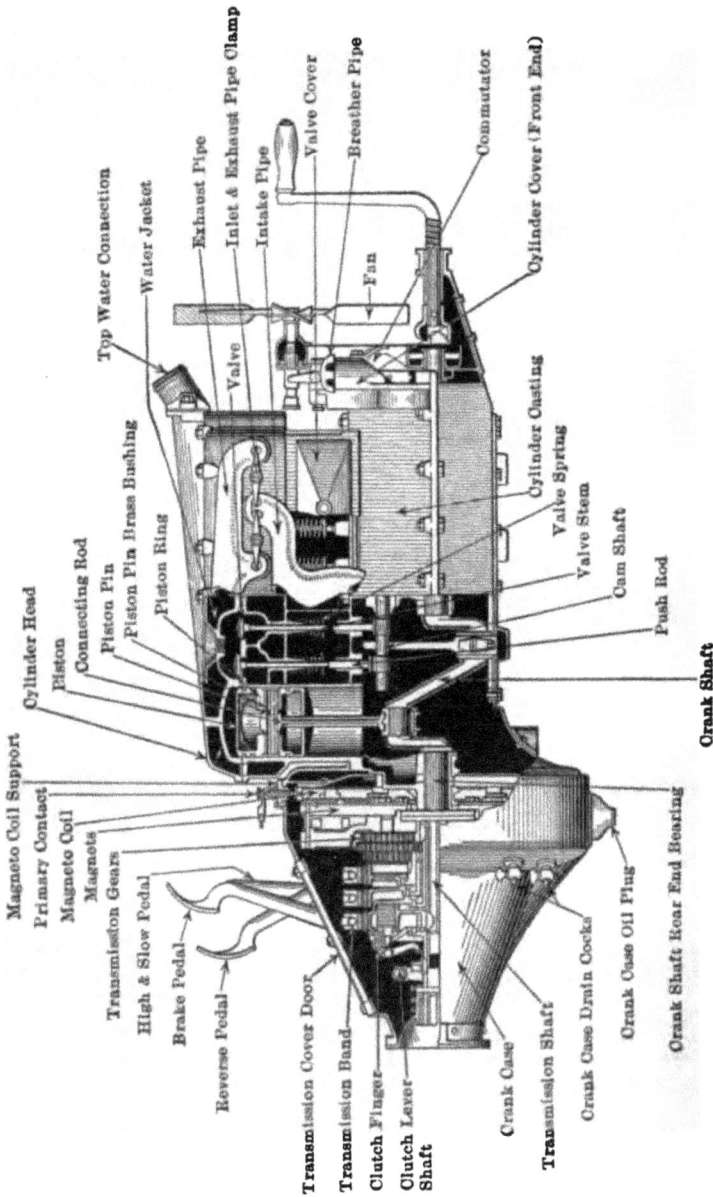

Fig. 11.—Part Sectional View of the Ford Four Cylinder Unit Power Plant Showing Important Parts of the Power Generating and Transmission System.

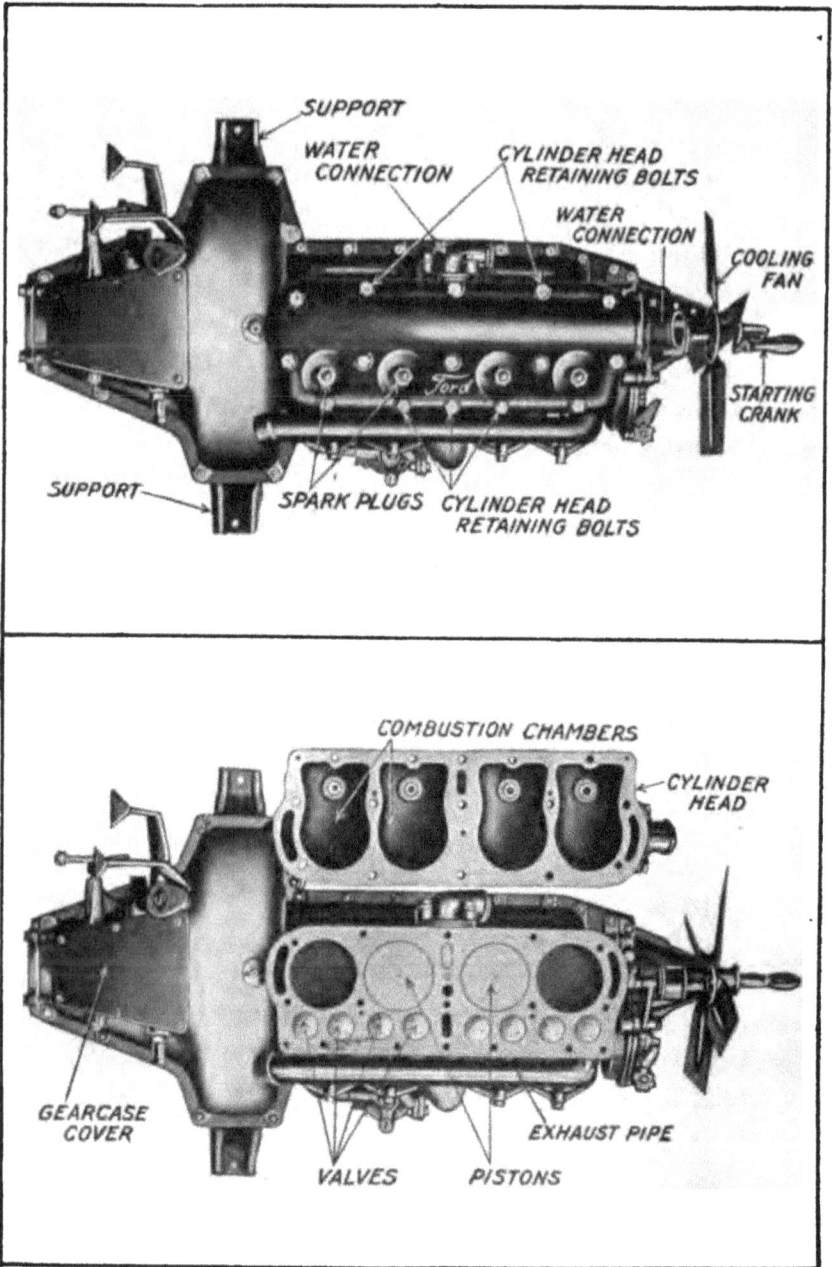

Fig. 12.—Depicting the Distinctive Design of the Ford Motor which Employs a Removable Cylinder Head to Permit Ready Access to the Combustion Chambers, Valves and Piston Parts.

Engine Parts and Their Functions. — A part sectional view of the Ford power plant is presented at Fig. 11, with practically all motor parts clearly outlined. The functions of the principal elements have been previously described, but in addition to these there are a number of minor parts which are equally necessary to secure efficient engine action. For example, each piston is provided with three packing rings fitting into grooves machined around each piston, two being above the piston pin and one below that member. The purpose of these packing rings is to prevent escape of the gas past the pistons on the compression or explosion strokes. It would not be possible to secure a practically gas tight joint without these rings because the piston could not be fitted so tightly to the cylinder that it would retain the gas. There would be considerable friction between the piston and cylinder walls, and in addition much power would be consumed because of this retarding influence which would increase as the piston became heated and expanded. By using packing rings, it is possible to make the piston enough smaller than the cylinder bore so that even when heated and expanded to its limit it will not seize or bind in the cylinder as long as lubrication is properly maintained. The piston rings are sufficiently elastic owing to a slotted opening in each ring to allow springing them over the piston when it is necessary to remove them for cleaning or examination. Another function of the piston ring is to take the wear caused to the rapid sliding movement of the piston. It is easier to renew relatively cheap piston rings instead of supplying new piston castings to compensate for wear after the engine has been in use for a time.

The connecting rods are attached to the pistons by the medium of a piston pin which is clamped in an eye at the upper end of the connecting rod and which oscillates in bronze bushings forced into the piston bosses. The connecting rod is a vanadium steel drop forging provided with a capped babbit bearing at its lower end, where it encircles the crank pin of the crank shaft. The cylinder unit consists of two members, an iron cylinder block which is cast integral with the upper half of the crank case and the cast iron cylinder head which is a removable member that can be taken off to provide easy access to the interior of the combustion chamber when carbon deposits must be cleaned out or the valves ground. The retaining bolts holding the cylinder head in place are clearly shown at the top of Fig. 12, while the appearance of the inside of the cylinder block when the cylinder head is removed is outlined in the lower portion of the illustration. Two valves are provided for each cylinder, these being side by side, all on the same side of the cylinder. The one that admits the fresh charge from the carburettor is called the "intake," whereas the one through which the exploded gas is driven out and which opens the passage between the combustion chamber and the exhaust manifold is called the "exhaust" valve.

The sectional view of the cylinder at Fig. 13, C, shows very clearly the manner in which the valve is installed, no valve seating spring being shown in order to simplify the illustration. Both inlet and exhaust valves are kept seated by means of coil springs which bear against a cap carried at the lower end of the valve stem. The appearance of the valve and its operating plunger is clearly shown at A, Fig. 13. The valves are opened at the proper time by the action of a simple member called a cam. This is of the form shown at Fig. 13, B, and consists of an approximately circular member except for a raised portion at one point on its periphery. One cam is provided for each valve, there being eight cams in the Ford engine all formed integrally with the cam shaft. As will be apparent by referring to B, Fig. 13, the valve plunger is raised when the pointed portion of the cam passes under it.

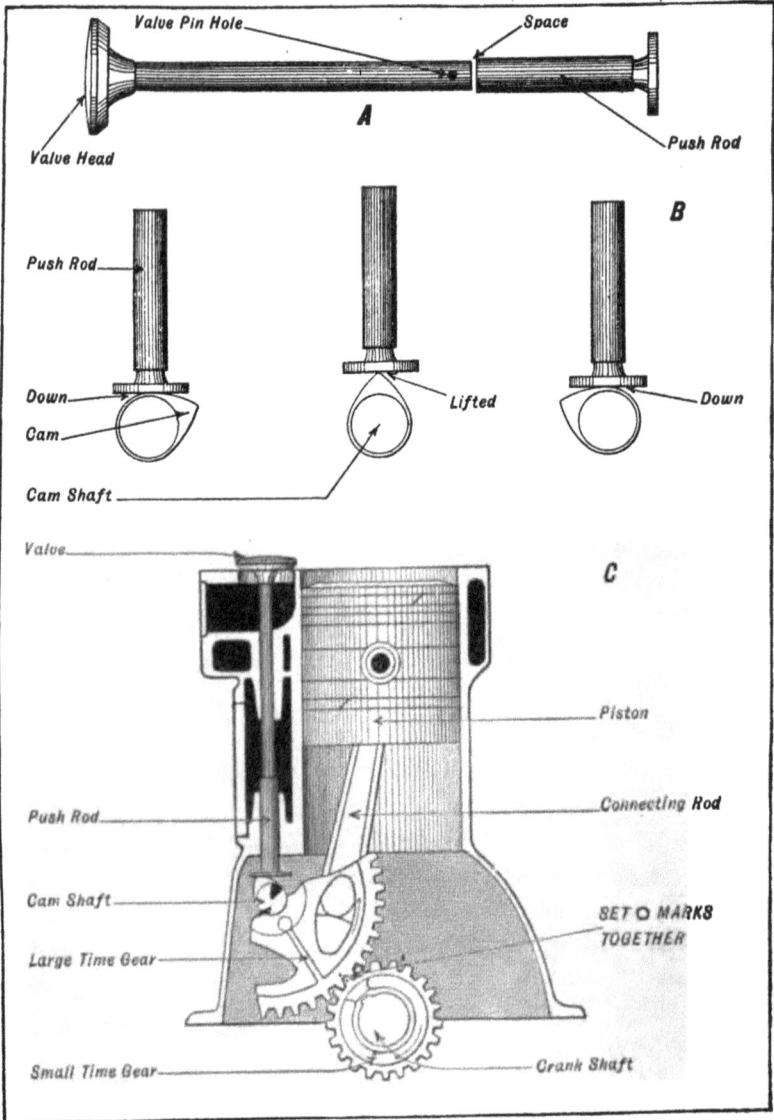

Fig. 13.—Showing Method of Ford Valve Construction and Operation.

The cam shaft is revolved at half the crank shaft speed, and is driven by means of a large gear which meshes with a pinion half its size called the small time gear keyed to the crank shaft. An important point to observe is that the valves are properly timed. This may be easily determined by checking to see if the factory timing has been disturbed. The timing is easily accomplished by having the tooth on the small time gear marked with a 0 fit between two teeth on the large timing gear at a corresponding 0 point. At this time the first cam on the cam shaft should point in a direction opposite from the zero marks, as shown at Fig. 13, C. The points of opening and closing of the valves and the troubles that are apt to materialize in these members will be discussed fully in the chapter on Engine Maintenance.

The crank shaft is one of the most important members of the engine, because it receives all the power delivered to it by the pistons and conveys the energy to the transmission gearing from which the drive is taken to the rear wheels of the car. The crank shaft, which is a vanadium steel forging, is supported by three main bearings, one at each end and one at the center. The flywheel member attached to the crank shaft serves to equalize the power application and steady the action of the engine as it serves to keep the crank shaft moving the brief intervals of time between the ending of one explosion and the beginning of the next one. The manifolds are pipes used to convey the gas in or out of the cylinder. The intake pipe is an aluminum or cast iron member that conveys the fresh gas from the carburettor to the inlet valve chambers. The exhaust pipe directs the burnt gas to the muffler. The four-blade, belt-driven fan at the front end of the motor forms part of the engine-cooling system. The starting crank is provided so the engine can be set in motion by hand to move the pistons until an explosion is obtained when it is desired to start it. The various parts of the transmission gear and control assembly will be described in the next chapter. Access to the interior of the crank case is obtained by removing a pressed steel plate on the bottom of the engine. The valve springs may be reached by removing steel plates covering the chambers in which they are housed. The inlet and exhaust pipes are held in place by simple clamp or stirrup members while the cylinder head is secured to the cylinder casting by substantial removable retaining bolts. The lower part of the engine crank case, which is a steel stamping, is fastened to the upper portion by another series of removable machine bolts. If the illustration at Fig. 11 is studied carefully no difficulty should be experienced in identifying the various parts of the Ford engine and understanding the work they do.

The Ford Carburetion System. — There is no appliance that has more material value upon the efficiency of the internal combustion motor than the carburettor or vaporizer which supplies the explosive gas to the cylinders. It is only in recent years that engineers have realized the importance of using carburettors that are efficient and that are so strongly made that there will be little liability of derangement. As the power obtained from the gas engine depends upon the combustion of fuel in the cylinders, it is evident that if the gas supplied does not have the proper proportions of elements to insure rapid combustion the efficiency of the engine will be low. When a gas engine is used as a stationary installation it is possible to use ordinary illuminating or natural gas for fuel, but when this prime mover is applied to automobile or marine service it is evident that considerable difficulty would be experienced in carrying enough compressed coal gas to supply the engine for even a very short trip. Fortunately, the development of the internal combustion motor was not delayed by the lack of suitable fuel. Engineers were familiar with properties of certain liquids which gave off vapors that could be mixed with air to form an explosive gas which burned very well in the engine cylinders. A very small quantity of such liquids would suffice for very satisfactory periods of engine operation. The problem to be solved before these liquids could be applied in a practical manner was to evolve suitable apparatus for vaporizing them without waste. Among the liquids that can be combined with air and burned, gasoline is the most common and is the fuel utilized by the majority of internal combustion engines employed in self-propelled conveyances.

Principles of Carburetion Outlined. — The process of carburetion is combining the volatile vapors which evaporate from the hydrocarbon liquids with certain proportions of air to form an inflammable gas. The quantities of air needed vary with different liquids and some mixtures burn quicker than do other combinations of air and vapor. Combustion is simply burning and as we have seen, it may be rapid, moderate, or slow. Mixtures of gasoline and air bum quickly, in fact the combustion is so rapid that it is instantaneous and we obtain what is commonly termed an "explosion." Therefore the explosion of gas in the automobile engine cylinder which produces the power is due to a combination of chemical elements which produce heat. If the gasoline mixture is not properly proportioned the rate of burning will vary, and if the mixture is either too rich or too weak the energy of the explosion is reduced and the amount of power applied to the piston is decreased proportionately.

In determining the proper proportions of gasoline and air, one must take the chemical composition of gasoline into account. The ordinary liquid used for fuel is said to contain about eighty-four per cent, carbon and sixteen per cent, hydrogen. Air is composed of oxygen and nitrogen and the former has a great affinity or combining power with the two constituents of hydrocarbon liquids. Therefore, what we call an explosion is merely an indication that oxygen in the air has combined with the carbon and hydrogen of the gasoline.

In figuring the proper volume of air to mix with a given quantity of fuel one takes into account the fact that one pound of hydrogen requires eight pounds of oxygen to bum it, and one pound of carbon needs two and one third pounds of oxygen to insure its combustion. Air is composed of one part of oxygen to three and one half portions of nitrogen by weight. Therefore, for each pound of oxygen one needs to burn hydrogen or carbon four and one half pounds of air must be allowed. To insure combustion of one pound of gasoline which is composed of hydrogen and carbon we must furnish about ten pounds of air to burn the carbon and about six pounds of air to insure combustion of hydrogen, the other component of gasoline. This means that to burn one pound of gasoline one must provide about sixteen pounds of air.

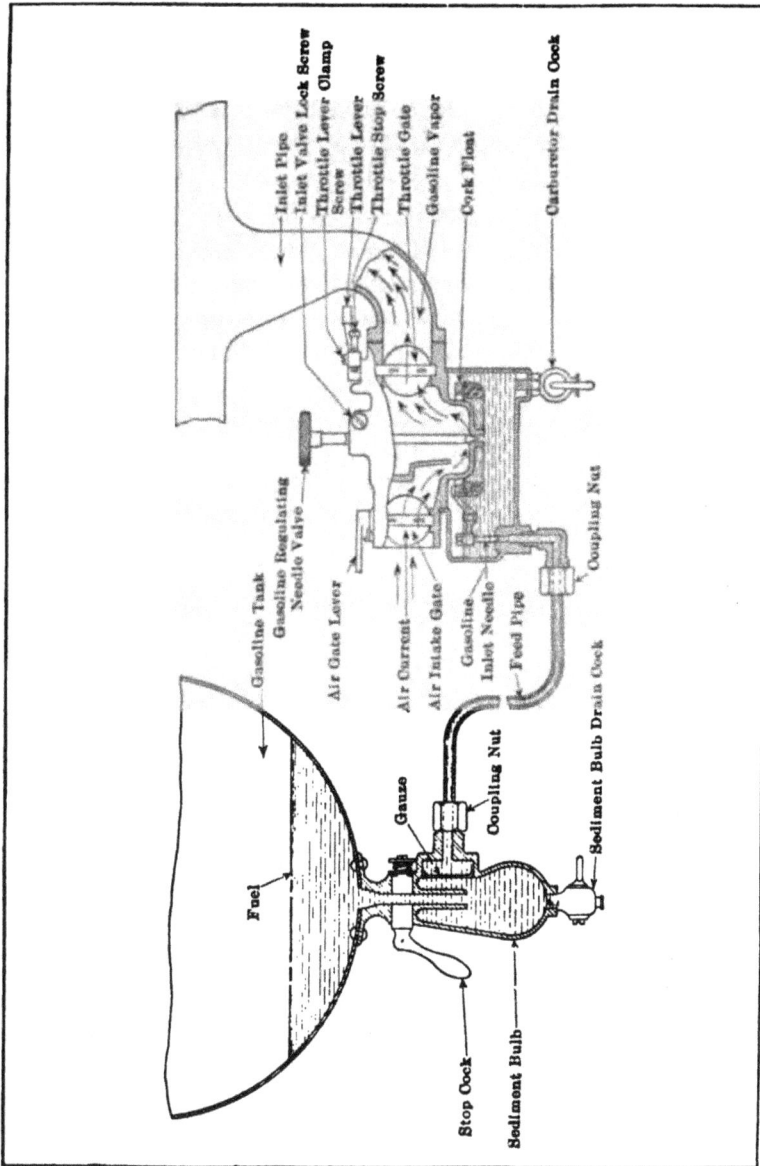

Fig. 14.—The Ford Model T Fuel Supply and Gas Making System.

While one does not usually consider air as having much weight, at a temperature of sixty-two degrees Fahrenheit about fourteen cubic feet of air will weigh a pound, and to burn a pound of gasoline one would require about two hundred cubic feet of air. This amount will provide for combustion theoretically, but it is common practice to allow twice this amount because the ele-

31

ment nitrogen, which is the main constituent of air, is an inert gas which instead of aiding combustion, acts as a deterrent of burning. In order to be explosive, gasoline vapor must be combined with certain quantities of air. Mixtures that are rich in gasoline ignite quicker than those which have more air, but these are only suitable when starting or when running slowly. The endeavor is to obtain a correct mixture of gasoline and air as it not only burns quicker but produces the most heat and the most effective pressure in pounds per square inch of piston top area.

The amount of compression of the charge before ignition also has material bearing on the force or power of the explosion. The higher the degree of compression the greater the force exerted and the more rapid the combustion of the gas. Mixtures varying from one part of gasoline vapor to four of air to others having one part of gasoline vapor to thirteen of air can be ignited, but the best results are obtained when the proportions are one to five or one to seven, as this mixture is the one that will produce the highest temperature the quickest explosion and the most pressure.

The Ford fuel system is clearly shown in accompanying diagram. Fig. 14. The gasoline supply is carried in a cylindrical galvanized iron tank under the front seat. This is joined to the carburettor by a simple pipe line of flexible soft copper tubing. The construction of one type of carburettor is shown in this illustration. Before the gasoline can flow to the vaporizer it must pass through the sediment bulb and filtering device on the bottom of the tank.

Utility of Gasoline Strainer. — Some carburettors include a filtering screen at the point where the liquid enters the float chamber in order to keep dirt or any other matter which may be present in the fuel from entering the float chamber. This is not general practice, however, and the majority of vaporizers do not include a filter in their construction. It is very desirable that the dirt should be kept out of the carburettor because it may get under the float controlled fuel inlet valve and cause flooding by keeping it raised from its seat. If dirt finds its way into the spraying orifice it may block the opening so that no gasoline will issue or may so constrict the passage that only very small quantities of fuel will be supplied the mixture. Where the carburettor itself is not provided with a filtering screen a simple filter is usually installed in the pipe line between the gasoline tank and the float chamber.

The simple form of filter and separator shown at Fig. 14 is used in the Ford fuel system. That illustrated consists of a simple hollow brass casting having a readily detachable gauze screen facing the outlet and a settling chamber of sufficient capacity to allow the foreign matter to settle to the bottom from which it may be drained out by a pet cock. Any water or dirt in the gasoline will settle to the bottom of the chamber, and as all fuel delivered to the carburettor must pass through the wire gauze screen it is not likely to contain impurities when it reaches the carburettor. The heavier particles, such as scale from the tank or dirt and even water, all of which have greater weight than the gasoline, will sink to the bottom of the chamber whereas light particles, such as lint, will be prevented from flowing into the carburettor by the filtering screen.

What a Carburettor Should Do. — While it is apparent that the chief function of a carburetting device is to mix hydrocarbon vapors with air to secure mixtures that will burn, there are a number of factors which must be considered before describing the principles of vaporizing devices. Almost any device which permits a current of air to pass over or through a volatile liquid will produce a gas which will explode when compressed and ignited in the motor cylinder. Modern carburettors are not only called upon to supply certain quantities of gas, but these must deliver a mixture to the cylinders that is accurately proportioned and which will be of proper composition at all engine speeds.

Flexible control of the engine is sought by altering the engine speed through regulation of the supply of gas to the cylinders. The power plant should run from its lowest to its highest speed without any irregularity in torque, i.e., the acceleration should be gradual rather than spasmodic. As the degree of compression will vary in value with the amount of throttle opening the conditions necessary to obtain maximum power differ with varying engine speeds. When the throttle is barely opened the engine speed is low and the gas must be richer in fuel than when the throttle is wide open and the engine speed high. When an engine is turning over slowly on low throttle the compression has low value and the

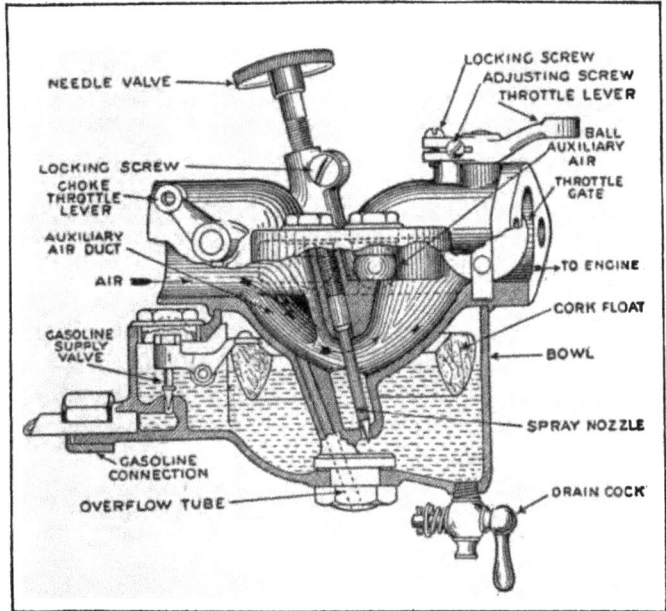

Fig. 15.—Part Sectional View of Special Kingston Carburetor Used on some Ford Model T Cars.

conditions are not so favorable to rapid combustion as when the compression is high. At high-engine speeds the gas velocity through the intake piping is higher than at low speeds, and regular engine action is not so apt to be disturbed by condensation of liquid fuel in the manifold due to excessively rich mixture or a superabundance of liquid in the stream of carburetted air.

The Ford Float Feed Carburettor. — The modem form of spraying carburettor is provided with two chambers, one a mixing chamber through which

33

the air stream passes and mixes with a gasoline spray, the other a float chamber in which a constant level of fuel is maintained by a simple mechanism. A nozzle or its equivalent is used in the mixing chamber to spray the fuel through and the object of the float is to maintain the fuel level to such a point that it will not overflow when the motor is not drawing in a charge of gas. Two different forms of carburettors have been used on Ford cars as regular equipment. One of these is shown in connection with the Fuel system at Fig. 14, the other is outlined at Fig. 15. The principle of action is the same for both types, except that one has an auxiliary air attachment consisting of a series of bronze halls which open progressively as the engine suction increases to admit more air into the mixture. This form is shown at Fig. 15. The Ford carburettor has but one adjustment and that by the gasoline needle valve. The fuel enters the float bowl through a connection at its side, its level in that member being regulated by the height of a cork float. The float raises as the supply increases to a point where the gasoline supply regulating valve is seated, this cutting off the flow of gasoline. As the gasoline is used up and the amount in the bowl becomes less the float lowers and through the medium of a bell crank or simple lever lifts the needle from its seat and permits more gasoline to flow into the float bowl from the gasoline tank. It is evident that a constant level of gasoline in the float bowl is maintained by the automatic action of the float-controlled needle. The amount of gasoline entering into the mixture is governed by the needle valve controlling the orifice through which the fuel flows from the float bowl to the interior of the mixing chamber. The volume of gaseous mixture entering the intake pipe, which in turn determines power and speed of the motor, is controlled by opening or closing a simple gate or shutter valve, similar to the damper used in a stove pipe, according to the engine speed desired by the driver.

A mixture that contains too much air and not enough gasoline is known as "Mean" mixture. If there is too much liquid fuel and not enough air the gas is called a "rich" mixture. Neither one of these conditions is desirable as the engine will be hard to start and lack power on a lean mixture and it will tend to overheat and be wasteful of fuel as well as promoting carbon deposits if the mixture is too rich. A rich mixture is indicated by a heavy black exhaust smoke having a disagreeable smell. When this condition is manifested the needle valve regulator on the dash should be screwed down or to the right until the engine begins to misfire, then the gasoline feed is gradually increased by opening the needle valve in the other direction slowly to that point where the motor runs steadily and at a high rate of speed with a full throttle opening, at the same time, there being no evidence of smoke in the exhaust. The reader should be cautioned that a white smoke coming out of the exhaust indicates too much lubricating oil and not too much gasoline. If popping sounds are heard in the carburettor when the engine is running it is because the mixture is too lean and the gasoline needle valve should be opened just enough to permit the engine to run well and yet not back fire.

As will be evident by studying the sectional views of Figs. 14 and 15 the gasoline level in the carburettor is at just such a height that a small pool of gasoline will collect at the bottom of the mixing chamber. The entering air current passing the air intake shutter is deflected toward the bottom of the float chamber by a suitable baffle plate or other obstruction and must sweep across the surface of the gasoline and become thoroughly impregnated with particles of liquid fuel before it can pass into the intake manifold. The action at high speed is different from that present at low speed because the greater engine suction and more rapidly moving air column does not give the gasoline a chance to accumulate in a pool at the bottom of the mixing chamber, but draws it from the orifice regulated by the needle valve in the form of a spray. The construction of a Ford carburettor is such that a correct gas is provided for easy starting while a leaner mixture is obtained as the engine speed increases and conditions become such that it can be utilized to advantage.

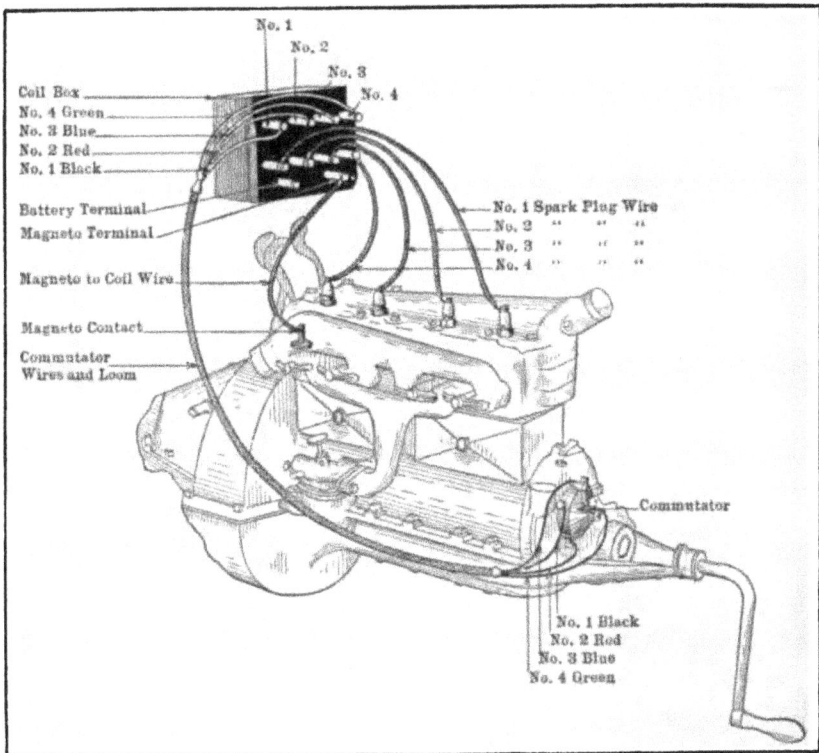

Fig. 16:—View of Ford Power Plant Showing Main Parts of the Ford Ignition System. Note Location of Timer and Induction Coil Box.

The Ford Ignition System. — The essential elements of any electrical ignition System, either high or low tension are: First, a simple and practical method of current production; second, suitable timing apparatus to cause the spark to occur at the right point in the cycle of engine action; third, suitable wiring and other apparatus to convey the current produced by the generator to the sparking member in the cylinder. The ignition system used on the Ford is a very simple and practical one. A four contact primary timer is mounted at the front end of the cam shaft, one contact being provided for each cylinder. "Wires run from this device to the coil units on the 3ash as shown in diagrams, Figs. 16 and 18. Other wires run from the induction coil units to the spark plugs, these are called secondary wires, while those going to the timer are called primary wires. When the car is shipped from the factory, no batteries are furnished, so only the magneto terminal is joined to the coil. This leaves one terminal free to be connected to a battery of dry cells when that is furnished by the owner. The magneto furnishes all the current that is needed to run the car and the engine may be easily started on the magneto when the coil vibrators are properly adjusted as this device supplies a strong current at ordinary cranking speeds.

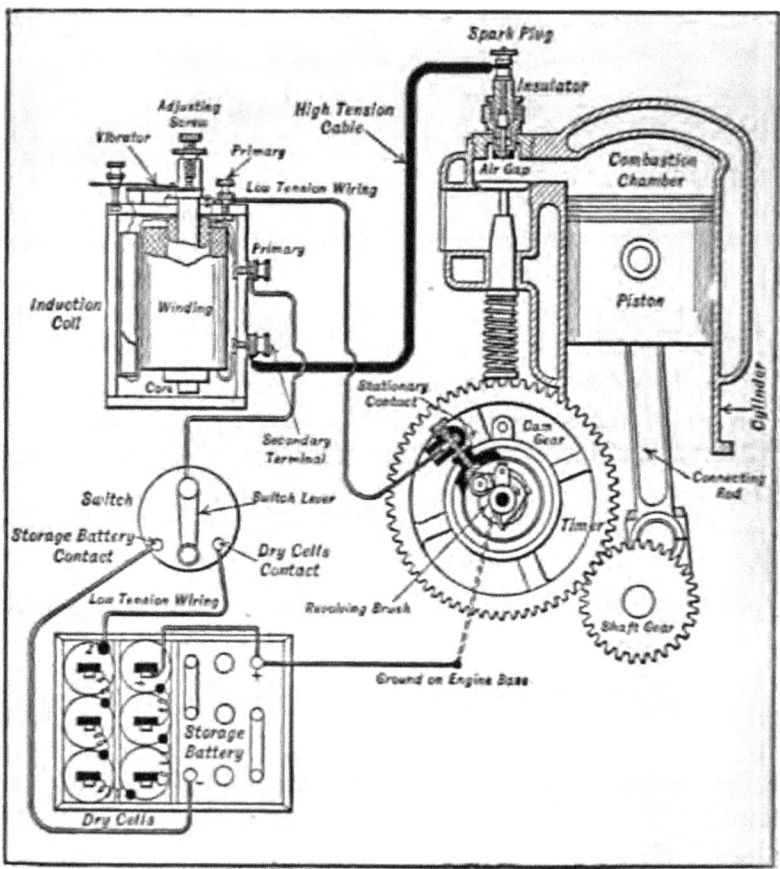

Fig. 17.—Diagram of Simple High Tension Ignition System for One Cylinder Motor to show Arrangement and Wiring of the Principal Parts.

Induction Coil System Explained. — In order to enable the reader to understand the basic principles of ignition apparatus the important parts of a simple battery ignition system for a one-cylinder engine are shown at Fig. 17. The current is supplied from two sources. One of these is a storage battery, the other a dry cell battery. A mechanical generator could be substituted for one of the batteries, if desired. The induction coil or transformer coil is utilized to intensify the low-tension current produced by the battery to one of greater value having sufficient voltage to jump the air gap between the points of the spark plug. The induction coil unit consists of a double wound coil surrounding a soft iron core piece. The primary coil or the one through which the battery current flows consists of two or three layers of comparatively coarse wire wound around the central core. The secondary coil is composed of a large number of turns of fine threadlike wire. Every time a current of electricity is permitted to flow through the primary coil it energizes the soft iron core, turns it into a magnet and the result is the production of a current in a secondary coil by magnetic induction, which has many times the voltage or pressure of the primary battery current producing it. though its amperage is greatly reduced. The average secondary coil used for ignition purposes will deliver a current of eight to ten thousand volts pressure. The amperage or quantity flowing is so slight that this voltage may pass through the human body without producing any injury.

Each time that a spark is desired between the points of the spark plug which projects into the interior of the combustion chamber, a contact is established between the revolving brush and the stationary contact of the timer. This permits the current to flow from either the dry or storage battery depending upon the position of the controlling switch lever, this current passing through the primary winding of the transformer coil. There is no electrical connection between the primary and secondary windings. The secondary winding is grounded at one end and is joined to the spark plug by a high tension cable at the other. In a simple one-cylinder engine of the four-cycle type but one explosion is possible in every two revolutions of the fly-wheel or in each four strokes of the piston. If the cam shaft which carries the revolving brush of the timer rotates at half the engine speed it will be apparent that but one electrical contact will be established for two revolutions of the crank shaft. This contact can be timed to take place only when the piston reaches the end of its compression stroke, at which time it is necessary to explode the gas to produce power.

A vibrator member, which is a simple automatic make-and-break arrangement operated by the magnetism of the induction coil core piece insures that the current will be sent through the primary winding in a series of waves during the main contact intervals. With a vibrator coil a stream of small sparks jump the air gap of the spark plug all the time that the revolving brush roller is in contact with the stationary contact segment of the timer. With a four cylinder engine it will be apparent that four contacts must be established for each two revolutions of the crank shaft and that the timer should have four stationary contact members which can be served by a common revolving brush. In the Ford ignition system four independent vibrator coils are provided, one for each cylinder. The method of wiring is clearly shown in Fig. 16, which gives the actual appearance of the parts of the ignition system while the course of the current may be readily followed by studying the diagram, Fig, 18. In this diagram the coil units are shown removed from the coil box in order to depict clearly the way the various connections are made. It will be evident that as soon as the revolving brush of the timer leaves the metal segment and all the time that it is in contact with the fibre ring in which the contact segments are imbedded that no current can flow through any of the induction coil units. As soon as the brush establishes contact with one of the segments the current is delivered to the unit that serves the cylinder that is about to fire, the wires being connected in such a way that coil unit supplying spark plug No. 1 works first, then No. 2, followed by No. 4 and lastly No. 3. This sequence of explosions is followed all the time the engine is in operation.

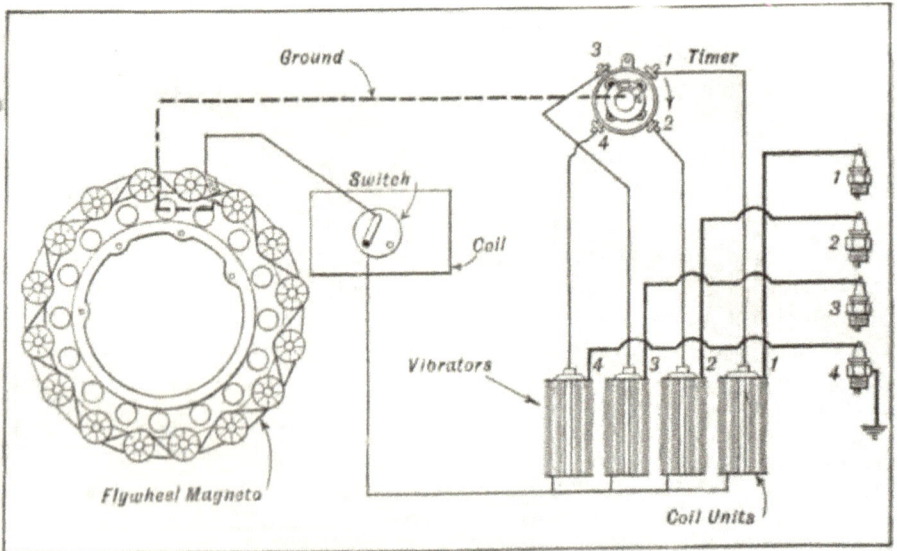

Fig. 18.—Wiring Diagram Showing Method of Connecting Parts of the Ford Ignition System.

The Ford Timer. — Anyone familiar with the basic principles of internal combustion engine action will recognize the need of incorporating some device in the ignition system, which will insure that the igniting spark will occur only in the cylinder that is ready to be fired and at the right time in the cycle of operations. There is a certain definite point at which the spark must take place, this having been determined to be at the end of the compression stroke, at which time the gas has been properly compacted and the piston is about to start returning to the bottom of the cylinder again. Timers or dis-

38

tributors are a form of mechanically operated switch designed so that hundreds of positive contacts which are necessary to close and open the circuit may be made per minute without failure. When a timer is to be used in connection with a four-cylinder engine the compact form shown at Fig. 19 is usually adopted. This has many desirable features and permits of timing the spark with great accuracy. The contact segments are spaced on quarters and are imbedded in a ring of fibre which is retained in a casing of aluminum. The central revolving element carries a lever which has a roll at one end and a tension spring designed to keep the roller in contact with the inner periphery of the fibre ring at the other. The segments are of steel and are accurately machined and hardened, and as the surface of the roller is also hardened, this form of timer is widely used because it provides a positive contact and works smoothly at all engine speeds, as well as having great endurance. Every time the roller makes contact with one of the segments, if the coil switch is on either battery or magneto, a current will flow from the generator through the timer and to the coil units to which the segment is wired. This produces a flow of current through the secondary wire to the spark plug where a spark jumps the air gap between the electrodes and explodes the compressed gas surrounding the plug points.

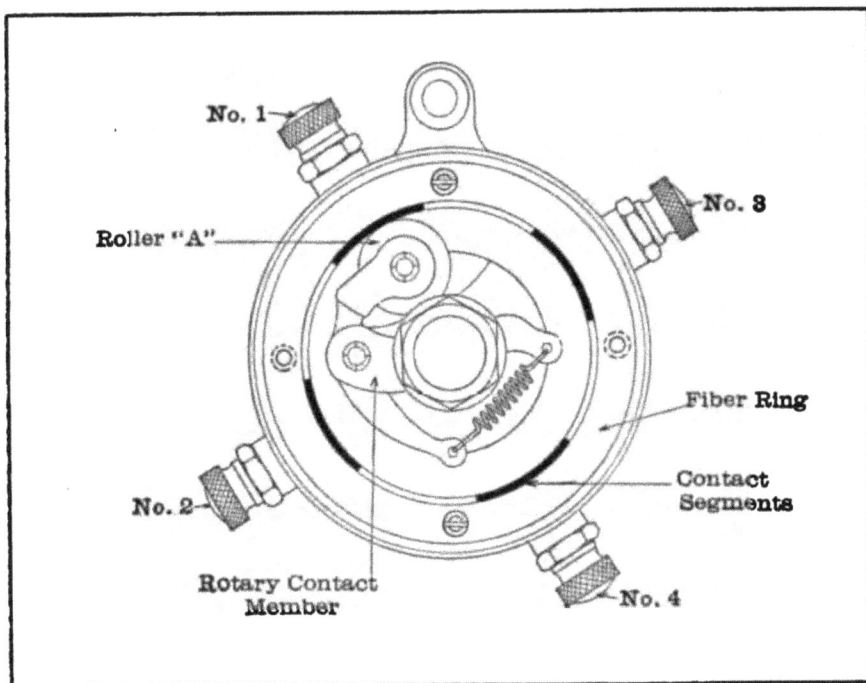

Fig. 19.—Parts of the Ford Ignition Timer.

Why a Magneto is Used on the Ford. — The fact that any chemical battery cannot maintain a constant supply of electricity has militated against their use to a certain extent and the modern motorist demands some appliance that will deliver an unfailing supply of electricity. The strength of batteries is reduced according to the amount of service they give. The more they are used the weaker they become. The modern multiple cylinder engines are especially severe in their demands upon the current producer and the rapid sequence of explosions in the Ford high speed four cylinder motor produces practically a steady drain upon the battery. When dry cells are used their discharge rate is very low and as they are designed only for intermittent work, when the conditions are such that a constant flow of current is required they are unsuitable and will soon deteriorate.

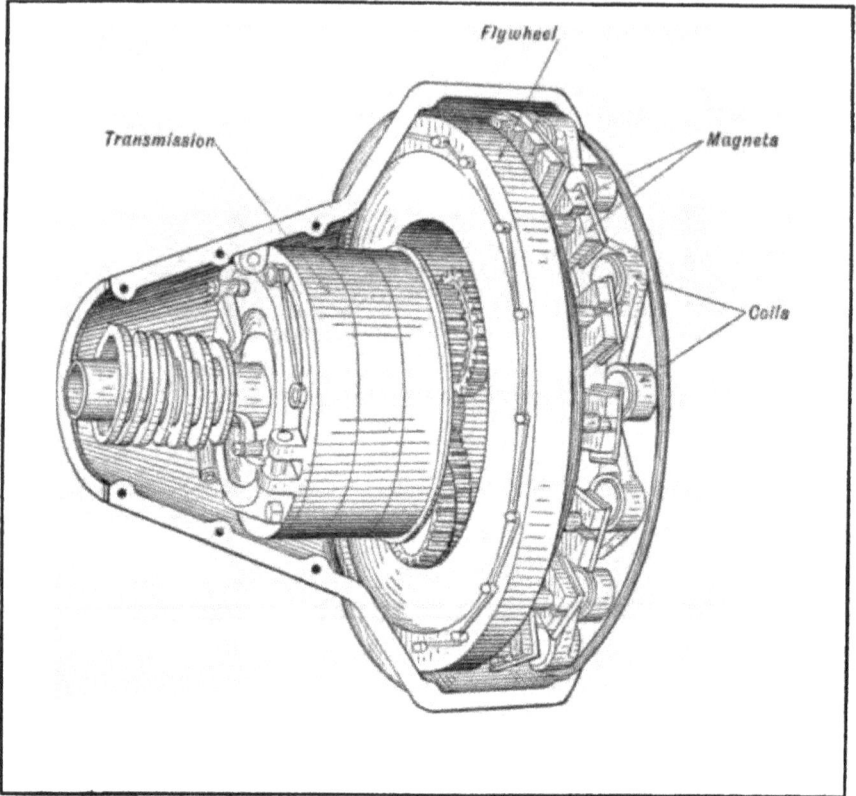

Fig. 20.—Showing Coils and Magnet that Comprise the Ford Magneto and their Relation to the Flywheel and Transmission Gear.

A very ingenious and practical application of the dynamo is shown at Fig. 20, this being used on the Ford car only. The electric generator is built in such a manner that it forms an integral part of the power plant. The magneto field is produced by a series of revolving permanent magnets which are joined to and turn with the flywheel of the motor. The sixteen current producing coils are carried by a fixed plate which is attached to the engine base. (See Fig. 21.) This apparatus is really a magneto having a revolving field and

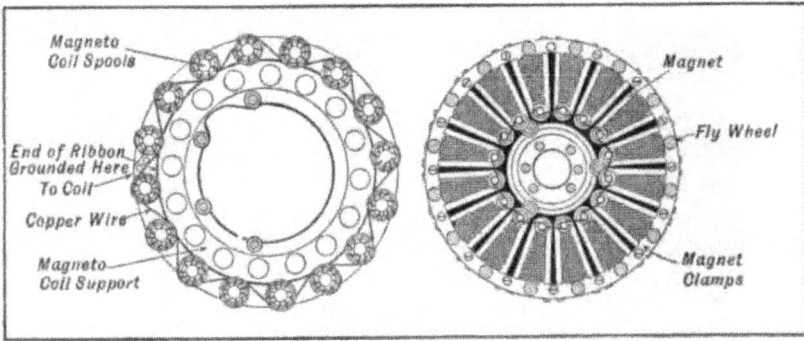

Fig. 21.—Views Showing Construction of Stationary Magneto Coil Carrying Member at Left and Rotary Magnet Carrier that Also Acts as the Motor Flywheel at Right.

a fixed armature, and as the magnets are driven from the flywheel there is no driving connection to get out of order and cause trouble. The coils in which the current is generated are stationary, no rotating commutators or fixed contact brushes are needed to collect the current because the electricity may be easily taken from the fixed coils by a simple direct connection. It has been advanced that this form of magneto is not as efficient as the conventional patterns because more metal and wire is needed to produce the current required. As the magnets which form the heaviest portion of the apparatus are joined to the flywheel, which can be correspondingly lighter, this disadvantage is not one that can "be considered seriously because the magnet weight is added to that of the motor flywheel, the combined mass of the two being equal to that of an ordinary balance member used on any other engine of equal power. The current supply will continue as long as the engine runs and a practically unfailing source of electricity is assured all Ford owners.

Many owners provide a set of dry cells as an auxiliary source of current, as these are of value in starting the motor sometimes under conditions where the engine cannot be cranked briskly enough to get a strong magneto spark. Dry cells are useful as a check upon the magneto and are also of value when adjusting the coil vibrators. The engine will also start on the spark sometimes without cranking when dry cells are used. The writer operated his car for two seasons with the magneto alone and without a battery and never felt the need of one. It was only when a storage battery was added to the equipment to operate electric side and tail lamps that the battery terminal of the coil was put into circuit, though it was seldom used.

Wiring Dry Cells. — One of the disadvantages of primary cells, as those types, which utilize zinc as a negative element are called, is that the chemical action produces deterioration and waste of material by oxidization. Dry cells are usually proportioned so the electrolyte and depolarizing materials become weaker as the zinc is used and when a dry cell is exhausted it is not profitable to attempt to recharge it because new ones can be obtained at a lower cost than the expense of renewing the worn elements would be. On

41

four-cylinder cars dry cells should be joined in multiple series, which is more enduring than if the same number were used independently in single series connection. A disadvantage of a dry cell battery is that it is suited only for intermittent service and it will soon become exhausted if used where the current demands are severe. For this reason, most automobiles in which batteries only are used for ignition, employ storage or secondary batteries to furnish the current regularly used and set of dry cells is provided for use only in cases of emergency when the storage battery becomes exhausted. To join dry cells in series, the zinc of one cell should be joined with the carbon of the adjacent member by a flexible conductor. This will leave the carbon of one end cell and the zinc of other end cell free so they can be joined to the apparatus in the outer circuit (see Fig. 22, A).

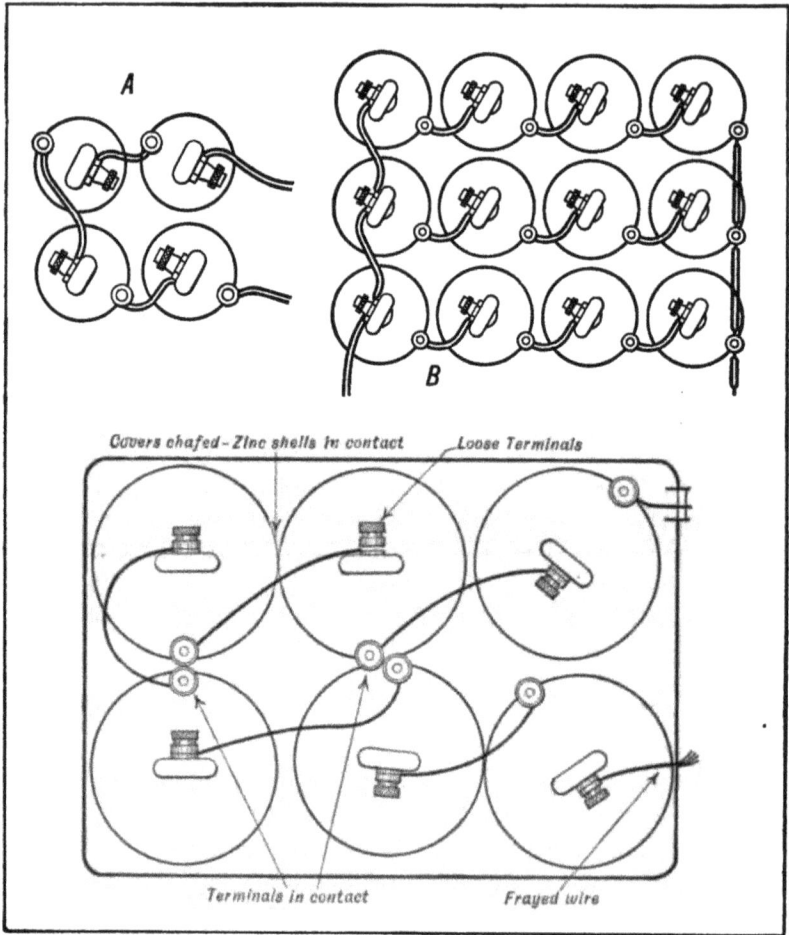

Fig. 22.—Illustrating Method of Connecting Dry Cells in Series at A and in Series Multiple at B. The Lower Illustration Shows Some of the Points to be Watched For When Dry Cells Are Installed in Metal Battery Boxes.

When it is desired to obtain more amperage or current quantity than could be obtained from a single cell they are joined in series-multiple connection, as at Fig. 22, B. With this method of wiring two or more sets of four cells which have been joined in series are used. The zinc of one set is joined with the zinc element of the others and the carbon terminals are similarly connected. Any number of sets of cells may be connected in series-multiple, and the amperage of the combination is increased proportionately to the number of sets joined together in this manner. When dry cells are connected in series the voltage of one cell is multiplied by the number of cells and the amperage obtained from the set is equal to that of one cell. When connected in series-multiple, as shown at Fig. 22, B, the amperage is equal to three cells, and the voltage produced is equivalent to that obtained from four cells. When twelve cells are joined in series-multiple the amperage is equal to that of one cell multiplied by three, while the voltage or current pressure is equal to that produced by one cell multiplied by the number of cells which are in series in any one set. By properly combining dry cells in this manner, batteries of any desired current strength may be obtained.

The terms "volt" and "ampere" are merely units by which current strength is gauged. The volt is the unit of pressure or potential which exists between the terminals of a circuit. The ampere is a measure of current quantity or flow and is independent of the pressure. One may have a current of high amperage at low potential or one having great pressure and but little amperage or current strength. Voltage is necessary to overcome resistance while the amperage available determines the heating value of the current. As the resistance to current flow increases the voltage must be augmented proportionally to overcome it. A current having strength of one ampere with a pressure of one volt is said to have a value of one watt, which is the unit by which the capacity of generators and the amount of current consumption of electrical apparatus is gauged.

The Master Vibrator System. — One of the most widely advertised accessories intended for the use of Ford car owners is called a "master vibrator"; this consists of a simple primary coil carrying a vibrator intended to serve all of the coil units, the regular vibrators with which these are provided being short circuited so that they do not operate. Opinions regarding the practical utility of a master vibrator differ greatly, some contending that it materially improves the steady operation of the engine while others do not believe that it is of any material benefit. The contention made by those favoring this device is that the use of one vibrator for all coil units provides a spark that will occur in each cylinder at exactly the same time in the cycle of operations, because it reduces the lag that might result from tardy action of one or more of the. individual unit coil vibrators. The argument of greater simplicity of having but one vibrator to adjust at any time is the more reasonable one. The writer did not find it necessary to use a master vibrator or any other of the legion of devices advertised to increase the efficiency of the Ford car. The vibrators of the four unit coil regularly provided gave very little trouble, as it was only the work of a few minutes to get all of these adjusted to the point where satisfactory engine action was obtained. The theory of irregular engine action due to lag of a poorly adjusted vibrator on any one of the coil units is only true in cases where the vibrator adjustment has been carried to a point where it is practically inoperative. The non-mechanical owner who cannot adjust the vibrators furnished on the regular Ford coil properly is not apt to have much success in adjusting that of a master vibrator, inasmuch as faulty adjustment of the one vibrator serving all coil units will throw the entire ignition system out of order, whereas if only one of the four coil units is not properly adjusted the engine will be able to run with some degree of power on the other three units.

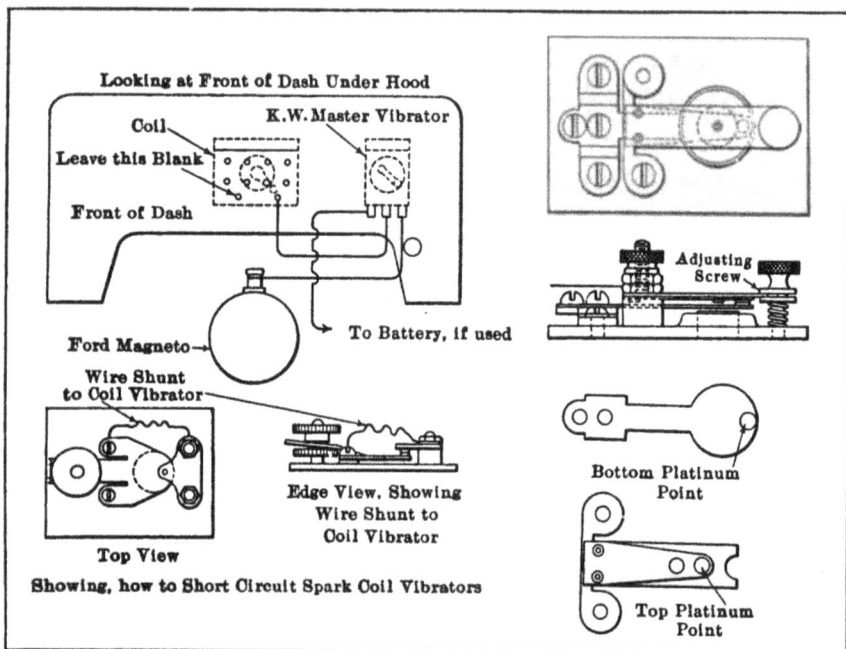

Fig. 23.—Showing Application of Master Vibrator in Ford Ignition System.

As many Ford cars have been fitted with a master vibrator by their owners, these afterward passing into other hands, it may be well for the reader to familiarize himself with the method of installation of this component. The diagram shown in the upper left corner of Fig. 23 shows the way the master vibrator is placed in circuit. The plan and side views below the wiring diagram show the simple method of short circuiting the regular spark coil vibrator by joining the vibrating and fixed portions with a short length of wire. At the right of the illustration are views showing the construction of the master vibrator, depicting the large platinum points necessary when one member serves four coil units, also the means provided for adjustment.

44

Why Cooling Systems Are Necessary. — The reader should now understand that the power of any internal combustion motor is obtained by the rapid combustion and consequent expansion of some inflammable gas. The operation in brief is that when air or any other gas or vapor is heated, it will expand, and that if this gas is confined in a space which will not permit expansion, pressure will be exerted against all sides of the containing chamber. The more a gas is heated the more pressure it will exert upon the walls of the combustion chamber by which it is confined. Pressure in a gas may be created by increasing its temperature, and inversely heat may be created by pressure. When a gas is compressed its total volume is reduced and the temperature is augmented. The efficiency of any form of heat engine is determined by the power obtained from a certain fuel consumption. A definite amount of energy will be liberated in the form of heat when a given quantity of any fuel is burned. The efficiency of any heat engine is proportional to the power developed from a definite quantity of fuel with the least loss of thermal units. If the greater proportion of the heat units derived by burning the explosive mixture could be utilized in doing useful work, the efficiency of the gasoline engine would be much greater than that of any other form of power producer.

There is a great loss of heat from various causes, among which can be cited the reduction of pressure through cooling the motor and the loss of the heat through the exhaust valves when the burned gases are expelled from cylinder. The loss through the water jacket of the average automobile power plant is over 50% of the total fuel efficiency. This means that more than half of the heat units that should be available for power are absorbed and dissipated by the cooling water. Another 16% is lost through the exhaust valve, and but $33 1/3\%$ of heat units do useful work. The great loss of heat through the cooling systems cannot be avoided, as some method must be provided to keep the temperature of the engine within proper bounds. It is apparent that the rapid combustion and continued series of explosions would soon heat the metal portions of the engine to a red heat if some means were not taken to conduct much of this heat away. The high temperature of the parts would burn the lubricating oil, even that of the best quality, and the piston and rings would expand to such a degree, especially when deprived of oil, that they would seize in the cylinder. This would score the walls, and the friction which ensued would tend to bind the parts so tightly that the piston would stick, bearings would be burned out, valves would warp, and the engine would soon become inoperative.

The best temperature to secure efficient operation is one on which considerable difference of opinion exists among engineers. The fact that the efficiency of an engine is dependent upon the ratio of heat converted into work compared to that generated by the explosion of the gas is accepted fact. It is very important that the engine should not get too hot, and at the other hand it is equally vital that the cylinder be not robbed of too much heat. The object of cylinder cooling is to keep the temperature of the cylinder below the danger point, but at the same time to have it as high as possible to secure maximum power from the gas burned.

Cooling Systems Generally Applied. — There are two general systems of engine cooling in common use: that in which water is heated by the absorption of heat from the engine and then cooled by air, and the other method in which the air is directed onto flanged cylinders and absorbs the heat directly instead of through the medium of water. When the liquid is employed in cooling it is circulated through jackets which surround the cylinder casting, and the water may be kept in motion by two methods. The one sometimes favored is to use a positive circulating pump of some form which is driven by the engine to keep the water in motion. The other system is to utilize a natural principle that heated water is lighter than cold liquid, and that it will tend to rise to the top of the cylinder when it becomes heated to the proper temperature and cooler water takes its place at the bottom of the water jacket.

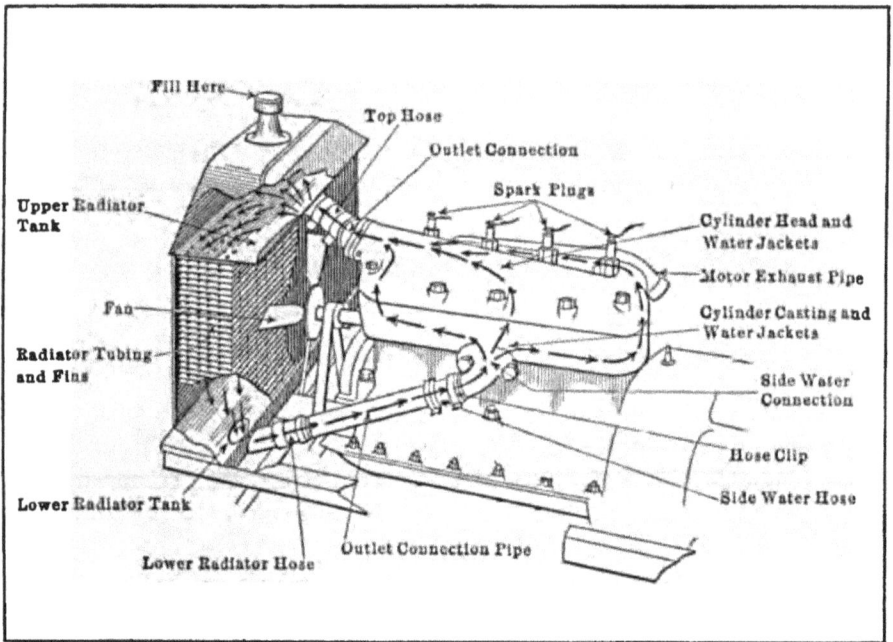

Fig. 24.—The Ford Thermo-Syphon Water Cooling System.

Ford Water Circulation by Natural System. — Some engineers contend that the rapid water circulation obtained by using a pump may cool the cylinders too much, and that the temperature of the engine may be reduced so much that the efficiency will be lessened. For this reason there is a growing tendency to use the natural method of water circulation as the cooling liquid is supplied to the cylinder jackets just below the boiling point, and the water issues from the jacket at the top of the cylinder after it had absorbed sufficient heat to raise it just about to the boiling point.

The Ford cooling system, depicted at Fig. 24, is very successful in practice, and is somewhat simpler than the forms in which a pump is used to maintain circulation. With this method, the fact that water becomes lighter as its temperature becomes higher is taken advantage of in securing circulation around the cylinders. The top of the water jacket of the block cast cylinder head is attached to the top center of the radiator, while the pipe leading from the bottom of that member is connected to a manifold which supplies cool water to the bottom of the cylinder jacket.

With a thermo-siphon system it is imperative that the radiator be carried at such a height that the cool water will flow to the water spaces around the cylinder by gravity. As the water becomes heated by contact with the hot cylinder and combustion chamber walls it rises to the top of the cylinders, flows to the cooler, where enough of the heat is absorbed to cause it to become sensibly greater in weight. As the water becomes cooler it falls to the bottom of the radiator, and it is again supplied to the water jacket. The circulation is entirely automatic and continues as long as there is a difference in temperature between the liquid in the cooler and that in the jacket. The circulation becomes brisker as the engine becomes hotter, and thus the temperature of the cylinders is kept more nearly to a fixed point. With the thermo-siphon system the cooling liquid is nearly always at its boiling point, whereas if the circulation is maintained by a pump the engine will become cooler at high speed and will heat up more at low speed. So long as the proper quantities of clean water are used in the radiator there is nothing that can interfere with proper engine cooling. There is no pump drive to complicate the construction and demand attention. It is an ideal cooling system for a car designed for use by the masses.

The radiator cools the water by dividing it into a number of fine streams as the liquid passes from the upper portion to the lower tank through a large number of fine copper pipes which are cooled by air currents passing over flanges soldered to them. A belt-driven fan is placed back of the radiator to insure a constant passage of air through the passages between the radiator tubes at all times that the engine is running. When the car is in operation, the air currents are increased in value by added air movement due to natural draft.

Theory of Lubrication. — The reason a lubricant is supplied to bearing points will be easily understood if one considers that these elastic substances flow between the close-fitting bearing surfaces, and by filling up the minute depressions in the surfaces and covering the high spots, act as a cushion which absorbs the heat generated, and takes the wear instead of the metallic surfaces. The closer the parts fit together the more fluid the lubricant must be to pass between their surfaces and at the same time it must possess sufficient body so that it will not be entirely forced out by the pressure existing between the parts. Oils should have good adhesive, as well as cohesive, qualities. The former are necessary so that the oil film will cling well to the surfaces of the bearings; the latter, so the oil particles will cling together and resist the tendency to separation which exists all the time the bearings are in operation.

When used for gas engine lubrication, the oil should be capable of withstanding considerable heat in order that it will not be vaporized by the hot portions of the cylinder. It should have sufficient cold test so that it will remain fluid and flow readily at low temperature. Lubricant should be free from acid or alkalies, which tend to produce a chemical action with metals and result in corrosion of the parts to which they are applied.

It is imperative that the oil be exactly the proper quality and nature for the purpose intended, and that it be applied in a positive manner. The requirements may be briefly summarized as follows: First — It must have sufficient body to prevent seizing of the parts to which it is applied and between which it is depended upon to maintain an elastic film, and yet it must not have too much viscosity in order to minimize the internal or fluid friction which exists between the particles of the lubricant itself. Second — The lubricant must not coagulate or gum, must not injure the parts to which it is applied, either by chemical action or by producing injurious deposits, and it should not evaporate readily. Third — The character of the work will demand that the oil should not vaporize when heated moderately, or thicken to such a point that it will not flow readily when cold. Fourth — The oil must be free from acid, alkalies, animal or vegetable fillers or other injurious agencies. Fifth — It must be carefully selected for the work required and should be a good conductor of heat.

Derivation of Lubricants. — The first oils which were used for lubricating machinery were obtained from animal and vegetable sources, though at the present time most of them are of mineral derivation. Lubricants may exist as fluids, semi-fluids, or solids. The viscosity will vary from spindle or dynamo oils, which have but little more body than kerosene, to the heaviest greases and tallows. The most common solid employed as a lubricant is graphite, sometimes termed "plumbago" or "black lead." This substance is of mineral derivation. Soapstone is also a lubricant, and is used in tires. The disadvantage of oil of organic origin, such as those obtained from animal fats or vegetable substances, is that they will absorb oxygen from the atmosphere which causes them to thicken or become rancid. Such oils have a very poor cold test, as they solidify at comparatively high temperatures and their flashing points are so low that they cannot be used at points where much heat exists. In most animal oils various acids are present in greater or less quantities, and for this reason they are not well adapted for lubricating metallic surfaces which may be raised high enough in temperature to cause decomposition of the oils.

Lubricants derived from the crude petroleum are called "Oleonaphthas," and they are a product of the process of refining petroleum through which gasoline and kerosene are obtained. They are of lower cost than vegetable or animal oils, and as they are of non-organic origin they do not become rancid or gummy by constant exposure to the air, and they will have no corrosive effect on metals because they contain no deleterious substances in their chemical composition. By the process of fractional distillation mineral oils of all grades can be obtained. They have a lower cold and higher flash test, and there is not the liability of spontaneous combustion that exists with animal oils.

The importance of minimizing friction at the various bearing surfaces of machines to secure mechanical efficiency is fully recognized by all mechanics, and proper lubricity of all parts of the mechanism is a very essential factor upon which the durability and successful operation of the motor car power plant depends. All of the moving members of the engine which are in contact with other portions, whether the motion is continuous or intermittent, of high or low velocity or of rectilinear or continued rotary nature, should be provided with an adequate supply of oil. No other assemblage of mechanism is operated under conditions which are so much to its disadvantage as the motor car, and the tendency is toward a simplification of oiling methods so that the supply will be ample and automatically applied to the points needing it.

In all machinery in motion the members which are in contact have a tendency to stick to each other, and the very minute projections which exist on even the smoothest of surfaces would adhere to each other if the surfaces were not kept apart by some elastic and unctuous substance. This will flow or spread out over the surfaces and smooth out the inequalities existing which tend to produce heat and retard motion of the pieces relative to each other.

How Ford Power Plant Is Lubricated. — The system of lubrication employed in the Ford power plant is an exceptionally simple one, requiring no apparatus other than that regularly forming a part of the engine. The construction of the magneto has been previously described, and mention made of the way the magnets are attached to the flywheel rim. These magnets also serve as a portion of the lubrication system being employed to circulate the oil. If one will refer to the part sectional view at Fig. 11, it will be apparent that a series of troughs are placed on the center line of each cylinder in the bottom plate, these being so arranged that as the connecting rods rotate the big ends dip into the troughs and scoop out some of the oil present in these members, throwing it about the engine interior and lubricating all parts exposed to the spray. It will be evident that all internal parts of the engine will be oiled continuously if some means is provided for keeping these troughs or channels full of lubricant.

This object is attained in a very simple manner by filling the flywheel compartment of the engine crank case to a definite height which is indicated by small drain cocks placed on the back side of the lower crank case compartment. This level is sufficiently high so the magnets are partially submerged in the oil as the flywheel revolves. It will be apparent that considerable oil will be scooped up by the projecting magnets, and these are utilized to lift oil into a small funnel attached to the side of the crank case and in the path of the oil stream. This funnel communicates with a brass tube that conveys the stream of lubricating oil to the front crank case compartment housing the timing gears. From this point the oil drains back, filling the troughs until they overflow, the surplus then flowing back into the flywheel compartment of the crank case. This system of lubrication also provides for thorough lubricity of the exposed planetary transmission gears carried in the gear case, which really forms the rear part of the engine crank case.

The oil is introduced into the engine through an opening obtained by removing the brass cover of the breather pipe. When the Ford engine is new and all crank case joints are tight so there is no leakage, the oil consumption will be equivalent to about one quart per hundred miles of car operation. The makers advise keeping the oil level at a point about midway between the two petcocks, but how this can be determined without the use of the X-ray can only be conjectured. They advise that carrying the oil level above the top petcock will result in excessive use of lubricant, whereas having the level below the lower petcock will be apt to result injuriously, owing to lack of lubrication. However, it is better to use too much oil than not enough, so most Ford owners fill the flywheel compartment to the height indicated by the top drain cock. Simple glass gauge fittings may be procured from accessory dealers by which the height of oil may be accurately gauged. These replace the lower petcock, and many Ford owners find it desirable to purchase this inexpensive fitting, as the level of the oil may be determined at a glance. During practically all the time that the writer had his car in operation, it was his rule to supply one quart of oil through the breather pipe for every five gallons of gasoline that was placed into the fuel tank. Then, the top petcock was opened until the surplus lubricant had drained out. With the Ford system of lubrication it is necessary to remove the crank case oil plug at the bottom of the flywheel compartment and drain out the old oil at least every five hundred miles, flushing out the interior of the crank case thoroughly with gasoline or kerosene and introducing enough lubricant after the oil plug had been replaced to bring the level to the proper height.

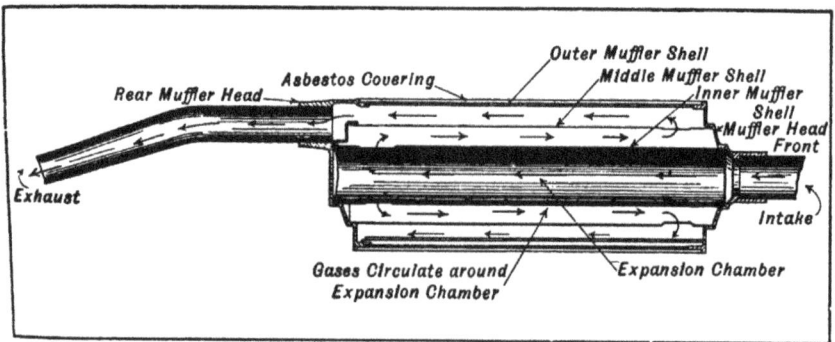

Fig. 25.—Sectional View Defining Construction and Method of Operation of the Ford Exhaust Gas Silencer.

as the gears are always in mesh these members cannot be injured by careless shifting. Individual clutches are used for speed selection, and as the operation of the clutch occurs at the same time that the desired speed is selected, any of the various speed changes desired may be easily effected by manipulating a single hand lever or pedal.

The planetary gearing shown at Fig. 27 is that used in Ford automobiles, and its operation is as follows: This contains only spur pinions. The flywheel web, A, serves as pinion carrier and driving member, having three lateral studs secured into it which carry triple planetary pinions. Gear B is the driven member, being keyed to the hub clutch drum C, which in turn is secured to driven shaft D. By applying a brake band to drum E, gear F is held stationary, pinion G rolls on it, and the smaller pinion H causes gear B to turn slowly in the same direction as pinion carrier A. By applying a brake band to drum I, gear J is held stationary, pinion K rolls on it, and the larger pinion H turns gear B slowly in the reverse direction. For the high gear, or direct drive, the friction clutch locks clutch drum C to the engine tail shaft, and the entire gear mechanism rotates as a unit. In this mechanism the master clutch, which provides the direct drive, is a multiple-disk form composed of two sets of steel disks, which are kept in contact and proper driving relation by means of a heavy coiled spring. The low and reverse speeds are obtained in the conventional manner by tightening the external contracting clutch bands, which are shown between the gearing and disk clutch in Fig. 28. One set of the high speed clutch plates drive

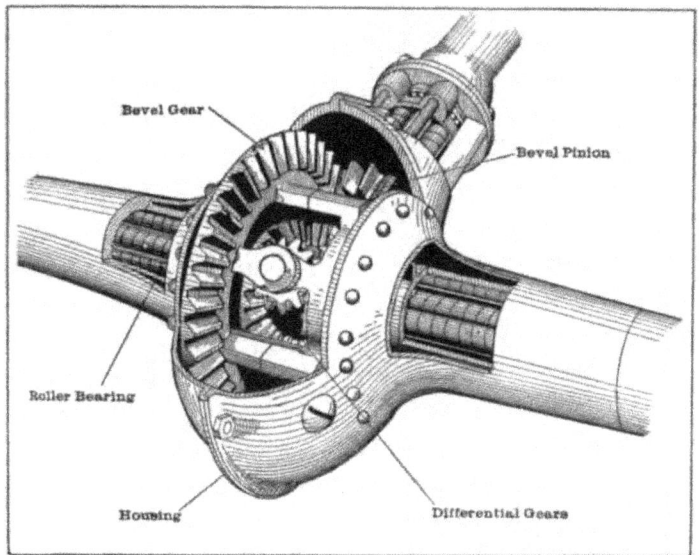

Fig. 29.—Cutaway View of the Ford Rear Axle Differential Housing Showing Arrangement of Bevel Driving Gearing and Differential Gears.

the drum C, and are driven by the other set which are keyed to the clutch disk carrier rotated by the engine crank shaft extension.

Planetary gearing has been very successful when properly designed and installed, and its chief disadvantage is that it is very difficult to provide more than two forward speeds and one reverse. For this reason it can only be adapted to light cars which have a surplus of power in the engine as the Ford.

The Ford Muffler. — When the exhaust gas of a gasoline engine is discharged into the open air directly from the valve ports, each discharge is accompanied by a sound resembling a gunshot. Evidently this would be very annoying to the public, so means are taken to silence the exhaust gases before they are discharged to the outer air. The Ford muffler, which is illustrated at Fig. 25, is a very simple assembly that silences the gas by permitting it to expand to a point where it is practically at atmospheric pressure before it is discharged to the air. The muffler is attached to one of the chassis side members, and is connected to the exhaust manifold attached to the cylinder casting by a piece of steel tubing. The muffler consists of two end castings having cylindrical ledges cast integrally which are used as supports for the concentric tubular members which divide the muffler into three distinct compartments. The gas from the exhaust pipe passes first into the central compartment, which is but slightly larger than the exhaust pipe. A number of passages or slots are pierced through the rear end of this chamber. The gas is discharged through the slots into the intermediate chamber, passing from this to the outer chamber through a series of openings at the front end of the middle muffler shell. The outer muffler shell serves as a casing for the assembly, and is covered with a sheet of asbestos which not only serves to muffle the sound made when the gas is discharged into the muffler, but which also serves to keep the heat of the muffler properly confined. The gas from the outer expansion chamber, which is formed by the space between the middle and outer muffler shells, issues to the air through a discharge pipe carried by the rear muffler head. The path of the gases is clearly shown by following the arrows through the various compartments. The Ford muffler is a very efficient one, reducing the sound of the exhaust to a point where it is not objectionable, yet at the same time not offering much back pressure to retard the free outflow of the gases. The silencing effect is obtained by breaking up the solid gas stream from the exhaust pipe into a number of smaller streams, and permitting these to expand in the concentric muffler chambers before they reach the air.

Chapter Three - Details of the Ford Chassis Parts

Next in importance to the power plant and its auxiliary groups are those chassis parts which have to do with the delivery of power from the engine crank shaft to the rear wheels. These parts are usually called the transmission members, and while they are not apt to give much trouble except to depreciate from natural wear as the car is used, it is well for the reader to become familiar with the method of operation and the relation the transmission parts bear to the other chassis components. The most important member, and one that is always in use, is the drive shaft which takes the engine power from the rear end of the transmission or change speed gearing to the bevel gears mounted in the rear axle which imparts motion to the wheels.

The parts comprising the transmission system are the clutch, the change speed gearing, the drive shaft, bevel driving gears, and the axle shafts which turn the wheels, these being carried by the rear construction. In considering the various parts, it will be well to define the reason why a clutch and change speed gearing are needed with a gasoline engine propelled automobile before describing the construction and operation of the Ford clutch and planetary gearing. In cars employing a sliding gear set the clutch is a separate member from the change speed gear, but in the Ford it forms an integral part of the mechanism depended on to obtain the two forward speeds and reverse ratios. The clutch and change speed gearing is mounted in an extension of the engine crank case, this insuring absolute alignment with the engine crank shaft.

Why Clutch Is Necessary. — In order to secure a better understanding of the general requirements of clutching devices, it will be well to consider the conditions which make their use imperative when an automobile is propelled by a hydrocarbon motor. If either a steam engine or an electric motor are installed as prime movers, it is not necessary to include any clutching device or gear set between them and the driving wheels, and these members may be driven directly from the power plant, if desired. With either of the forms mentioned the power is obtained from a separate source which may be disconnected from the motor by the simple movement of a throttle valve or switch lever. Steam or electric motors are also capable of delivering power in excess of their rating, and are more flexible than internal combustion power plants.

If steam is the motive agent it is generated and contained in a special device, as a boiler, and the amount of power delivered by the engine to which the boiler is connected will vary with the amount of steam admitted and its pressure. If the steam supply is interrupted entirely, the engine and the car which it drives is brought to a stop. When it is desired to start again, a simple movement of the throttle-valve lever will permit the steam to flow from the boiler to the engine cylinders again, and the vehicle is easily set in motion. If it is desired to reverse the car, the steam flow is reversed by a simple mechanical movement and the engine will run in the opposite direction to that which obtains when the car is driven in a forward direction.

If an electric motor drives a vehicle, the electrical energy is obtained from a group of storage batteries. When these are fully charged varying amounts of electric current may be drawn from them and allowed to flow through the windings of the field or armature of the motor and different ratios of power or speed obtained. The vehicle is easily started by completing the circuit between the motor and the source of current and stopped by interrupting the supply of electrical energy. As the flow of electricity can be reversed easily by a switch, the car may be driven backward or forward at will, and as the speed may be easily varied by changing the value of the current strength there is no need of speed changing or reversing gears.

When a gasoline engine is fitted, conditions are radically different than with either a steam or electric power plant. The power developed depends upon the number of explosions per unit time and the energy augments directly as the number of explosions and revolutions of the crank shaft increase up to a certain point. It is not possible to start a gasoline engine under full load because the power is obtained by the combustion of fuel directly in the cylinder, and as there is no external source of power to draw from, it is obvious that the energy derived depends upon the rapidity with which the explosions follow each other. It has been demonstrated that a certain cycle of operation is necessary to secure gasoline engine action, and it is imperative that the engine revolves freely until it attains sufficient speed to supply the torque or power needed to overcome the resistance that tends to prevent motion of the ear before it can be employed in driving the vehicle.

Then, again, it is very desirable that the vehicle be started or stopped independently of the engine. With a steam or electric motor the vehicle may be started just as soon as the driving power is admitted to the prime mover, but with a gasoline engine it is customary to interpose some device between the engine and the driving wheels which make it possible to couple the engine to the wheels or driving gearing and disconnect it at will. The simplest method of doing this is by means of some form of clutching device which will lock the rear wheel driving shaft to the crank shaft of the engine.

Clutch Forms and Their Requirements. — Clutch forms that have been applied to automobile propulsion are usually of the frictional type, though some have been devised which depend upon hydraulic, pneumatic, or magnetic energy. Those which utilize the driving properties of frictional adhesion are most common, and have proven to be the most satisfactory in practical application. The most important requirement in a clutch is that this device be capable of transmitting the maximum power of the engine to which it is fitted without any power loss due to slipping. A clutch must be easy to operate and but minimum exertion should be required of the operator. When the clutch takes hold, the engine power should be transmitted to the gears and driving wheels in a gradual and uniform manner or the resulting shock may seriously injure the mechanism. When released it is imperative that the two portions of the clutch disengage positively, so that there will be no continued rotation of the parts after the clutch is disengaged. The design should be carefully considered with a view of providing as much friction surface as possible to prevent excessive slipping and loss of power. It is very desirable to have a clutch that will be absolutely silent whether engaged or disengaged. If the clutch parts are located in an accessible manner it may be easily removed for inspection, cleaning, or repairs. It is desirable that adjustment be provided, so a certain amount of wear can be compensated for without expensive replacement. A simple, substantial design, with but few operating parts, is more to be desired than a more complex device which may have a few minor advantages, but which is more likely to cause trouble.

The friction clutch in its various efficient types is the one that more nearly realizes the requirements of the ideal clutch. As a result this form is now universally recognized by automobile designers, and all standard gasoline automobiles utilize some form of friction clutch which is included with the planetary speed reduction gearing on the Ford car. These devices are capable of transmitting any amount of power if properly proportioned, and permit of gradual engagement and positive disconnection. Most friction clutches are simple in form, easily understood, and may be kept in adjustment and repair without difficulty.

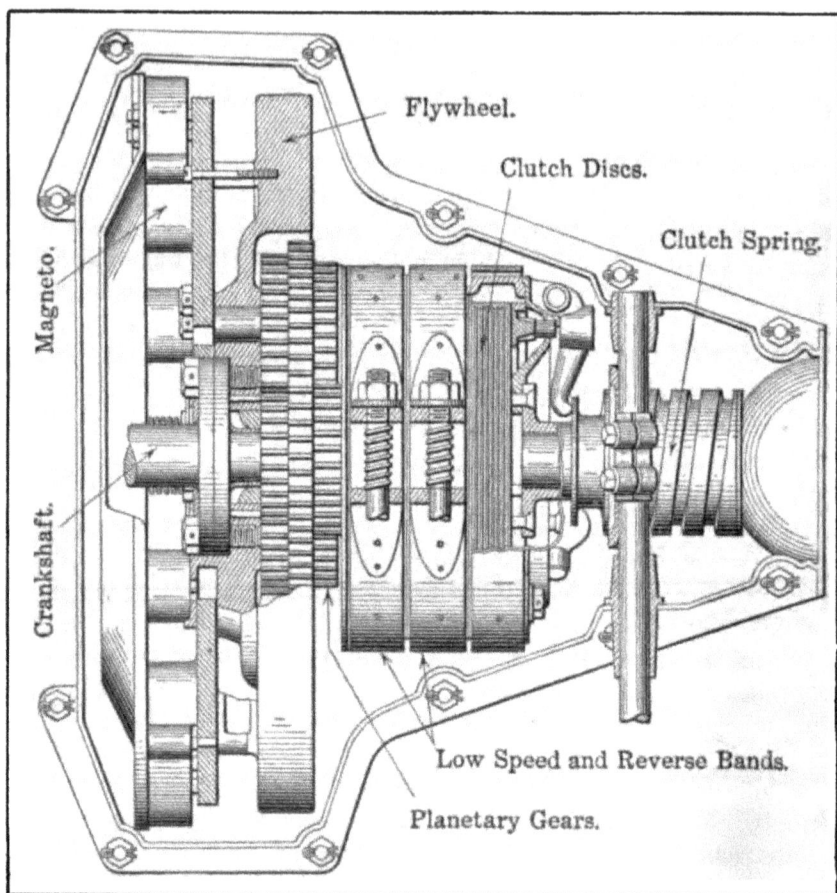

Fig. 26.—Plan View of the Ford Planetary Gearing Showing Method of Carrying Triple Planetary Spur Pinion Assemblies and Actuating the High Speed Disc Clutch Assembly.

How Friction Clutches Transmit Power. — To illustrate the transmission of power by the frictional adhesion of substances with each other we can assume a simple case of two metal disks or plates in contact, the pressure existing between the surfaces being due to the weight of one member bearing upon the other. If the disks are not too heavy, it will be found comparatively easy to turn one upon the other, but if weights are added to the upper member, a more decided resistance will be felt which will increase directly as the weight on the top disk, and consequently the pressure between the disks, increases. It may be possible to add enough weight so it will be practically impossible to move one plate without turning the other. It is patent that if one of these plates was mounted rigidly on the engine shaft and one applied to the transmission shaft so that it had a certain amount of axial freedom and pressure of contact was maintained by a spring instead of weights, a combination capable of transmitting power would be obtained. The spring pressure applied to one disk would force it against the other, and one shaft could not turn without producing a corresponding movement of the other. The Ford clutch, shown at Fig. 26 is a multiple disk form.

Why Change Speed Gearing is Needed. — Those who are familiar with steam or electricity as sources of power for motor vehicles may not understand the necessity for the change speed gearing which is such an essential component of the automobile propelled by internal combustion motors. In explaining the reason for the use of the clutch it has been demonstrated that steam or electric motors are very flexible, and that their speed, and consequently the power derived from them, could be varied directly by regulating the amount of energy supplied from the steam boiler or the electric battery, as the case might be. If, for example, we compare the steam engine with the explosive type, it will be evident that the power is produced in the former by the pressure of steam admitted to the cylinders as well as the quantity and the speed of rotation. When the engine is running slowly and a certain amount of power is needed, more steam can be supplied the cylinder, and practically the same power obtained, as though the steam pressure was reduced and the engine speed increased. The internal combustion motor is flexible to a certain degree, providing that it is operating under conditions which are favorable to accelerating the motor speed by admitting more gas to the cylinders. There is a definite limit, however, to the power capacity or the effective pressure of the explosion, and beyond a certain point it is not possible to increase the power by supplying vapor having a higher pressure as is possible with a steam engine.

In an explosive motor we can increase the power after the maximum explosive pressure has been reached only by augmenting the number of revolutions. Whereas it is possible to couple a steam engine or an electric motor directly to the shafts driving the wheels, it is not possible to do this with gasoline engines, and some form of gearing must be introduced between the motor and the driving wheels in order that the speed of one relative to the other may be changed, as desired, and the engine crank shaft turned at speeds best adapted to produce the power required, and to allow the rear wheels to turn at speeds dictated by the condition of the roads or the gradients on which the car is operated. It is customary in all automobiles of the gasoline-burning type, where combustion takes place directly in the cylinders, to interpose change speed gearing which will give two or more ratios of speed between the engine and the road wheels. As it is not possible to reverse the automobile engine utilized in conventional cars, it is necessary to add a set of gears to the gear set to give the wheels a reverse motion when it is desired to back it.

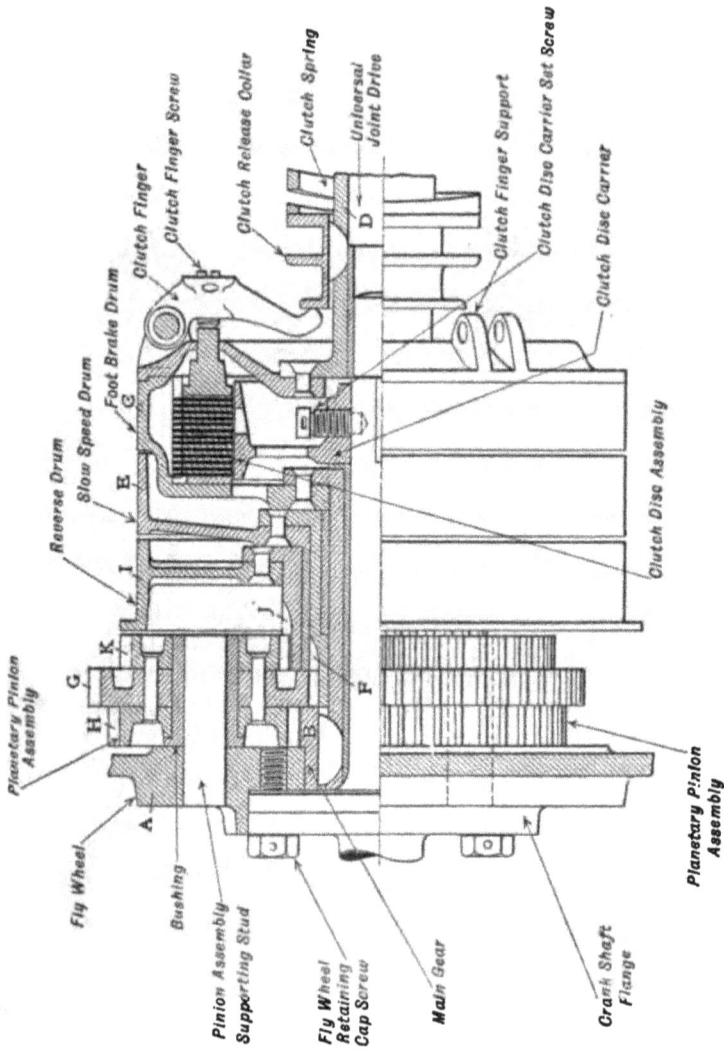

Fig. 27.—Part Sectional View of the Ford Planetary Gearing Showing the Relation of the Planetary Pinion Assembly, the Reverse, Slow Speed and Foot Brake Drums and the Clutch Disc Assembly.

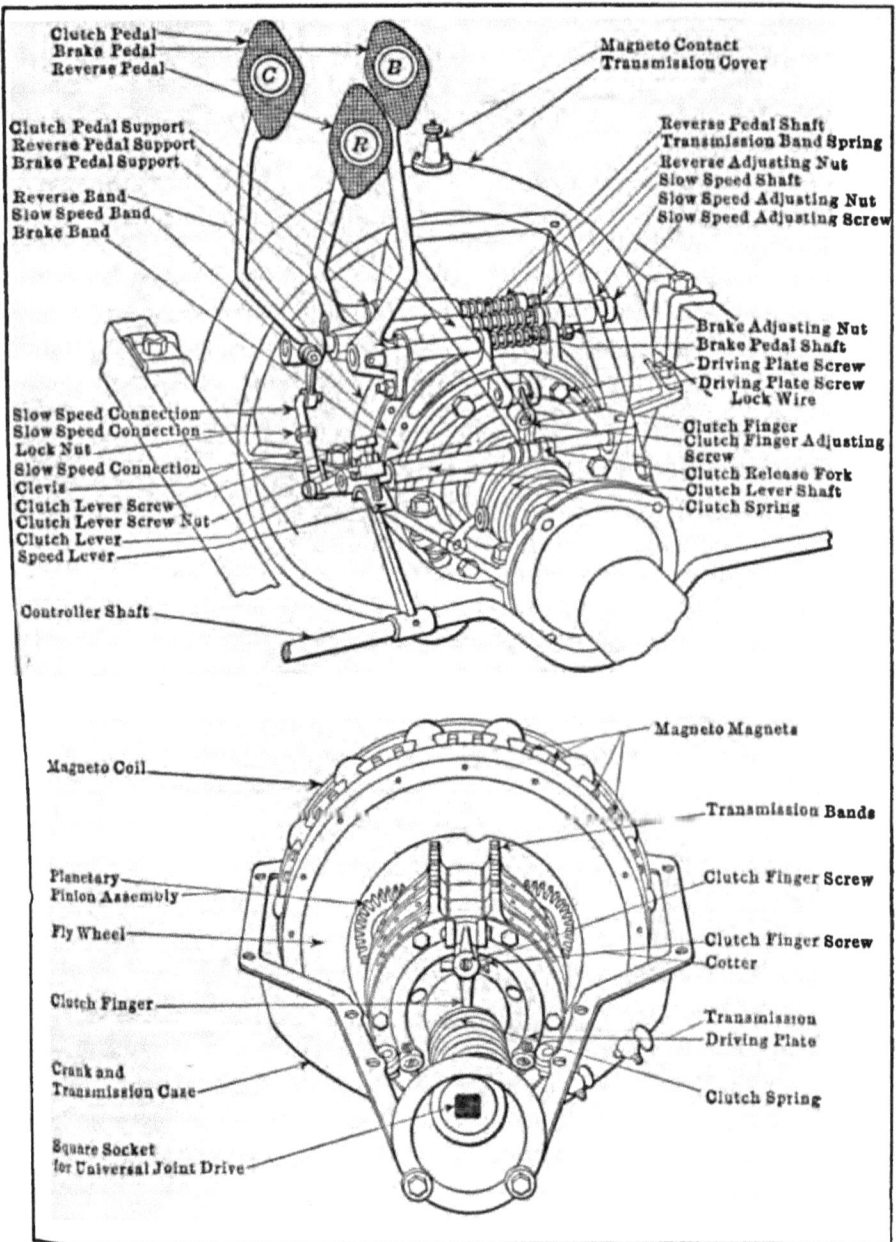

Top diagram labels (left side):
Clutch Pedal
Brake Pedal
Reverse Pedal

Clutch Pedal Support
Reverse Pedal Support
Brake Pedal Support

Reverse Band
Slow Speed Band
Brake Band

Slow Speed Connection
Slow Speed Connection Lock Nut
Slow Speed Connection Clevis
Clutch Lever Screw
Clutch Lever Screw Nut
Clutch Lever
Speed Lever

Controller Shaft

Top diagram labels (right side):
Magneto Contact
Transmission Cover

Reverse Pedal Shaft
Transmission Band Spring
Reverse Adjusting Nut
Slow Speed Shaft
Slow Speed Adjusting Nut
Slow Speed Adjusting Screw

Brake Adjusting Nut
Brake Pedal Shaft
Driving Plate Screw
Driving Plate Screw Lock Wire

Clutch Finger
Clutch Finger Adjusting Screw
Clutch Release Fork
Clutch Lever Shaft
Clutch Spring

Bottom diagram labels (left side):
Magneto Coil

Planetary Pinion Assembly

Fly Wheel

Clutch Finger

Crank and Transmission Case

Square Socket for Universal Joint Drive

Bottom diagram labels (right side):
Magneto Magnets

Transmission Bands

Clutch Finger Screw

Clutch Finger Screw Cotter

Transmission Driving Plate

Clutch Spring

Fig. 28.—Phantom View of the Ford Planetary Gearset Showing the Control Pedal Assembly at Top. View of Gearing Partially Disassembled Showing Brake Bands and Other Parts at the Bottom.

How Planetary Gearing Operates. — The planetary or epicycle transmission is an easily operated form of speed gear that has been very popular on small cars. This has many features of merit; it provides a positive drive, and

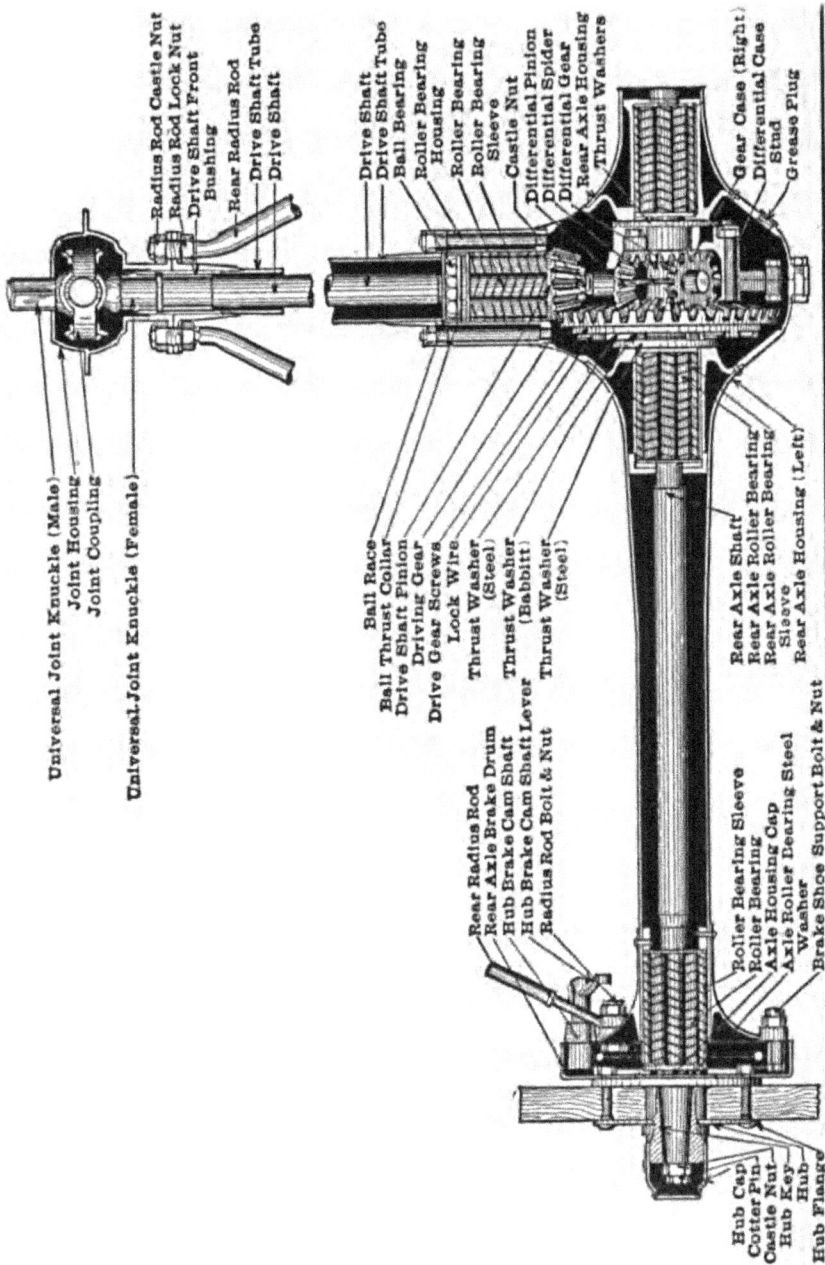

Fig. 30.—Sectional View of Ford Model T Rear Axle Showing Driving Gears, Differential, Power Transmission Shafts and Supporting Bearings.

While such gearing is not very efficient on low and reverse speeds as considerable power is absorbed in friction, on the high speed or direct drive it is superior to any other form of change-speed gearing because the entire assembly is locked to the crank shaft, no gears are turning idly and the weight of the gearing serves merely as an additional flywheel member. With light

59

cars like the Ford practically all roads may be negotiated on the direct drive and the low speed is very seldom used. Considerable trouble was experienced with the early forms because it was difficult to keep oil in the case, but in the Ford design special care has been taken in housing the reduction gears so these are constantly oiled, and both wear and noise, which were formerly detrimental to the adoption of this form of gearing and which militated largely against its general use, have been eliminated to a large extent.

Method of Power Transmission. — The power delivered to the gear set from the motor crank shaft is taken by means of the universal joint and propeller shaft to bevel gearing forming part of the rear axle. This driving gearing is shown at Fig. 29 while a sectional view showing the arrangement of the rear axle parts is presented at Fig. 30. The propeller shaft is enclosed in a drive shaft tube which also acts as a torque member to resist the tendency of the rear axle to rotate while the wheels are driving the car or when the hub brakes are applied to stop the car. This tube terminates in a universal joint casing made in the form of a ball with a portion cut off the top, this fitting in a suitable carrying member or ball seat, machined in the back end of the transmission case. The front end of the propeller shaft revolves in a plain bearing, while the rear end which carries the bevel driving pinion is supported by a flexible roller bearing. The bevel pinion meshes with a large bevel gear, often called the "ring gear," which is attached to the differential housing in the manner indicated at Fig. 29. A portion of the differential housing is cut away in this illustration to show the method of carrying the differential pinions and the way these are in mesh with the differential gears attached to the wheel driving axle shafts. Part of the axle housing is also cut away on each side to show the roller bearings which may be more clearly seen in the sectional view of entire rear construction at Fig. 30. When the motor is operating and the low or high speed is engaged, the universal joint and the shaft to which it is attached turn clockwise when viewed from the front of the car. In other words this shaft revolves in the same direction as the crank shaft of the motor though its speed of rotation depends on whether the low speed band or the high speed clutch is engaged.

With the low speed engaged the engine shaft is turning faster than the drive shaft, though with the high speed clutch transmitting the power the propeller shaft and engine crank shaft turn at the same speed. The power then goes through the universal joint and the shaft to which the bevel pinion is attached, from this to the large ring gear attached to the differential casing and from the differential mechanism to the axle shafts connected to the road wheels. When the engine is turning clockwise, the large gear turns forward, as do the axles and wheels attached to them, and as a result the car will move in a forward direction. When the reverse motion control pedal is depressed and the reverse friction band is constricted around the reverse drum of the gearing the gears previously described come into action, thus reversing the motion of the universal joint and pinion drive shaft causing the large gear in the rear axle to turn in a direction opposite to that obtained when either the high or low speeds are in action and thus producing a backward motion of the car.

It will be apparent that there is considerable difference in the size of the drive gears, the bevel pinion being much smaller than the ring gear attached to the differential. This is done because the ring gear must turn slower than the engine crank shaft, as it would not be practical to rotate the road wheels at a speed equal to that of the engine crank shaft because the resistance to car movement is too great to be overcome by such a direct application of power. The bevel pinion is provided with eleven teeth and the large gear it drives has forty teeth. Consequently, the driving shaft and its pinion will make 3%i revolutions for every one of the large gear when the high speed clutch or direct drive is engaged. When the low speed is brought into action the reduction is considerably increased by the gears in the transmission. In this case, the engine crank shaft will make about ten revolutions to one of the rear wheels.

One of the important elements of the driving system is the universal joint which is a flexible, though positive driving coupling that permits rotating the driving shaft even though that member is at an angle with the engine crank shaft. This slope is made necessary because the differential and drive gearing in the rear axle are carried lower than the gear set, so it will be apparent that it is necessary to provide some driving coupling that is capable of compensating for this lack of alignment. The universal joint is composed of three main parts, two knuckle joints and a joint ring. The ring is made in two half sections riveted together which serve as a bearing for the male and female knuckle joint driving pins. As will be noted by consulting Fig. 30, the male knuckle joint is so called because it has a squared end intended to slip into a square machined in the transmission shaft. The boss of the female knuckle joint is in the form of a sleeve formed to slip over the end of the driving shaft and secured thereto by a pin passing through both. The square end of the male joint may move back and forth in the transmission shaft to compensate for any slight end movement that may be present as the universal joint ring member rocks on the knuckle joint pins. The universal joint is housed in the globular member indicated, which is filled with lubricant to provide for thorough oiling of the moving parts.

The rear construction, as the entire rear axle assembly is called, supports the rear end of the chassis and in turn is supported by the road wheels. These are members somewhat similar in design to certain forms of carriage wheels, the wooden spokes being mounted between hub^ flanges at the central part of the wheel and forced into a wooden felloe band to which the tire-carrying rim isi attached at the outer ends. The rear wheel flanges are of metal and they are securely attached to a central hub member which is bored tapering to fit the tapered end of the drive axle. The axle is provided with a key which fits a keyway in the hub member and when the wheel hub is clamped on the axle taper by the retention nut the wheel cannot turn unless the axle turns with it. As a result when the energy of the motor is applied to the driving axles through the medium of the differential gear the road wheels must rotate with them. The wheel is prevented from backing off of the taper by a suitable clamp nut which in turn is locked in place by a split pin which passes through the axle and which fits into slots milled across the end of the nut. The rear wheel hubs carry a pressed steel brake drum which is retained by the same bolts holding the wheel assembly together. This drum serves to house the emergency brake shoes and their operating cams. The rear construction therefore consists of three casings or housing members, one serving to carry the propeller shaft while the other two are bolted together to form the housing for the axle shafts carrying the wheels and the differential and driving gear mechanism.

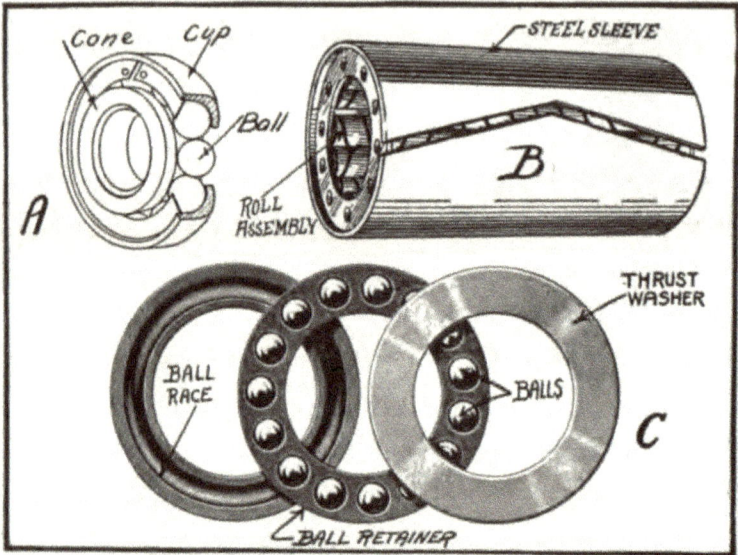

Fig. 31.—Types of Anti-Friction Bearings Used in the Ford Car. A—Cup and Cone Type Angular Contact Ball Bearings Similar to Those Used in the Front Wheels. B—Hyatt Flexible Roller Bearing. C—Special Ball Bearings for Resisting End Thrust Only.

The Ford Axle Bearings. — Careful study of the cutaway view of the differential housing at Fig. 29 and of the rear axle assembly at Fig. 30 will show that the various driving shafts are supported by anti-friction bearings at all points subjected to heavy loads. For example, the driving shaft is supported at the universal joint-end by a plain bushing which answers the purpose because it serves merely to guide the shaft and is not subject to any great stress. At the pinion end, however, the load is greater and a plain bearing would wear out very quickly, besides consuming a lot of power all the time it was in use. The pinion end of the driving shaft, therefore, is fitted with a large flexible roller bearing and a ball thrust bearing.

The function of this thrust bearing is to compensate for the tendency to end movement of the driving shaft which results because of the angularity of

62

the faces of the bevel driving gearing. When the motor is propelling the car, the driving reaction on the angular teeth of the ring gear produces a decided end thrust against the pinion shaft. The roller bearing which is utilized to take care of the radial load or to prevent side movement of the shaft is not capable of withstanding this end thrust so a special ball thrust bearing must be provided to assist the roller bearing to preserve the proper relation between the driving pinion and the ring gear. All wheels of the Ford car are carried by anti-friction bearings. The front wheels are mounted on cup and cone ball bearings of the general type shown at A, Fig. 31, their practical application being shown at Fig. 32. This form of bearing consists of a pressed steel cup member forced into the hub shell casting and a cone member fitting the axle spindle tightly. The space between the cup and cone is filled with steel balls which carry the load with a rolling motion and thus have much less friction than a plain bearing in which the surfaces must slide over each other.

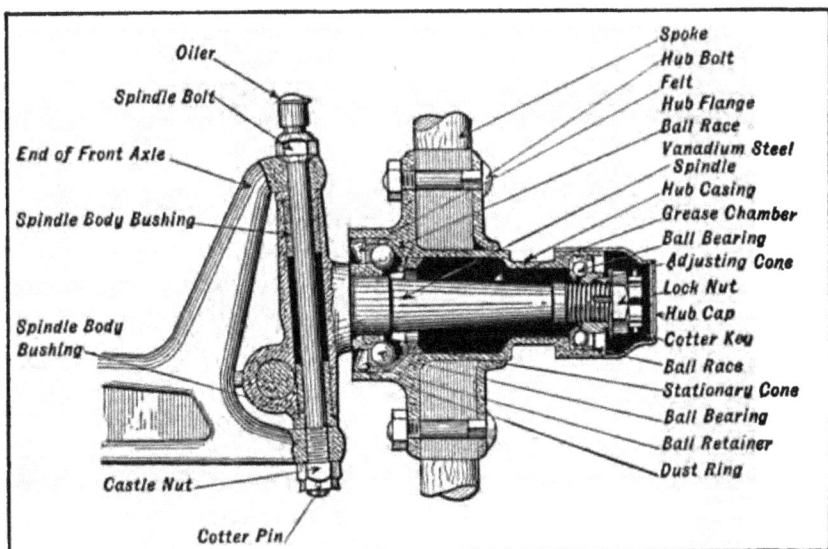

Fig. 32.—Sectional View of Ford Front Wheel Hub Showing Method of Installing Cup- and Cone- Type Ball Bearings.

The differential mechanism and the wheel end of the axle utilize roller bearings of the general form shown at Fig. 31, D. This bearing consists of a cage carrying a number of spiral rollers which roll on the steel shaft but which do not bear directly against the housing tube as a steel sleeve is introduced to form a track for the rollers to run on. The ball thrust bearing which is used at the front end of the roller bearing supporting the pinion end of the propeller shaft is of the general form shown at Fig. 31, C. In this bearing the raceways have grooved ball tracks formed on their faces, the balls being placed between them in such a way that the bearing is suitable only to take loads coming from a direction approximately parallel with the driving shaft.

These loads are called end thrusts while loads applied at right angles to the driving shafts, as, for example, the weight load on the roller bearings at the wheel end of the axle are termed radial loads. As a roller bearing consumes less power than a plain bearing and as they are more enduring and require less attention as well as being inexpensive, the entire axle and differential assembly is carried by roller bearings.

Each axle shaft revolves in two roller bearings, one being placed near each side of the differential and one at each of the outer ends of the drive axle housing near the wheel. The roller bearing consists of a group of hardened steel rollers in the form of close wound coil springs which are prevented from coming in contact with each other by a cage or retainer which is clearly shown at Fig. 31, B. It will be apparent that the outer bearings or those at the wheel end of the axle carry practically all of the radial load due to car weight and driving strains. To compensate for end thrust on the wheel drive axles, such as is present when the wheels skid, thrust bearings are provided at each side of the differential gear case. The axle shafts cannot move outward because they are securely keyed to the differential gear members, these transferring any of the thrust load to babbit metal rings carried at each side of the differential housing which in turn are sandwiched in between steel thrust plates interposed between the axle housing and the babbit ring on one side and the differential case and the babbit ring on its other side. The function of the radius rods which extend from the ends of the axle shaft housing to the flanged fitting back of the ball joint is to strengthen the entire rear axle construction as they form a triangle having the apex at the universal joint end. Any tendency of either end of the axle housing to move backwards or forwards or for the housing enclosing the driving shaft to twist or bend is corrected by these rods.

Purpose of Differential Gear. — One of the most important yet inconspicuous elements of any form of automobile driving system is the differential gear, but as this is usually placed at a point where it is not easily seen by the motorist and as but very little trouble is experienced from this mechanism, many owners of cars are not aware of its existence and do not realize the important work performed by this relatively simple component. Without a differential gear it would be difficult to control the machine when driving around corners, so this really performs an important function with both steering and driving systems.

When turning corners with any four-wheel vehicle the outer wheels must turn at a higher rate of speed than the inner ones because they are describing a larger arc of the circle. The more sharply the vehicle is turned the greater the difference in velocity between the inner and outer wheels. In a horse-drawn conveyance all the wheels are independent of each other and may all revolve at different speeds if necessary, without interfering with each other or impairing the action of the conveyance. In an automobile different conditions prevail because while the front wheels are usually independent of each other, the driving wheels must be connected together so that each will receive its share of the energy produced by the motor and will perform its quota of the work incidental to propelling the vehicle.

In order to permit one of the driving wheels to turn at a lower speed than its mate in rounding a corner the balance or differential gear is used. Its simplest application is shown at Fig. 33. From this it is patent that the driving axle is split in the center and that the wheels are mounted on and driven by distinct axle shafts. (See Fig. 30.) At the inner end of each shaft a bevel gear is carried, these being firmly secured to the axles so they revolve with them. The main bevel-driven gear, which is actuated by the driving pinion turned by the engine, is mounted independent of the axles and is coupled to them by means of small bevel pinions which are applied so that they will drive the gears on the axle shafts. Assuming that all the gears are in mesh, as outlined, and that power is being applied to the driven gear, and that the resistance to traction is the same at both rear wheels, the entire assembly comprised of driven gear, the differential pinions attached to it and the axle shafts revolve as a unit.

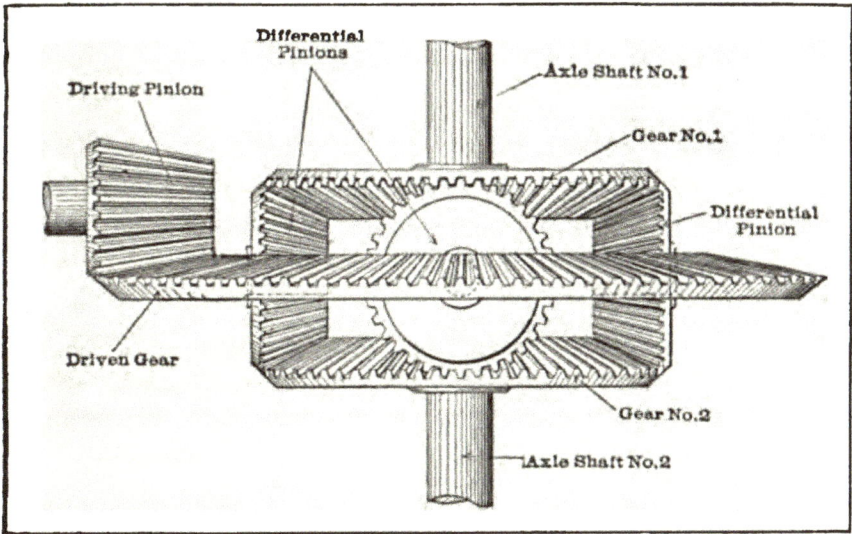

Fig. 33.—Simplified Diagram to Accompany Explanation of Differential Gear Action.

If the resistance against the driving wheels varies so one wheel tends to revolve faster than the other, the differential pinions will not only turn around on the studs on which they are mounted, but at the same time will run around the gears on the axle shafts, because the bevel driven gear carrying the studs on which the differential pinions revolve moves forward. When turning a corner the outer wheel must turn so much faster than the inner member that it is just the same as though one of the wheels was held stationary and the other turned. If both wheels are turning forward at the same speed, the differential pinions remain stationary and act simply as a lock which forms a driving connection between gear No. 1 on axle shaft No. 1 and gear No. 2 on axle shaft No. 2. This will mean that both wheels must turn in the same direction as long as the work is uniformly distributed. Just as soon as the resistances are unequal the differential pinions will turn on their supporting studs and one member may turn at comparatively slow speed while the other revolves at a much faster rate. The action of the differential pinions may be clearly understood by reference to Fig. 33 and giving due consideration to the following principles. The same resistance at the point of contact between the driving wheels and the ground prevents the pinions from revolving on their own studs, and in this case they are carried around by the supporting members and the ring gear. If the resistance upon axle shaft No. 1 is greater than that on axle shaft No. 2, the ring gear will rotate forward with the wheel offering the least resistance and the differential pinions will turn on their studs and run over the surface of the gear which tends to remain stationary, this being the one against which there is the greatest resistance. The differential pinions can thus turn independently of one gear wheel and run over its surface without turning it, and at the same time act as a clutching member of sufficient capacity on the other gear and axle to carry them in the same direction as the ring gear and at a ratio of speed which will depend upon the difference in resistance between the driving members and the ground.

Utility of Motor-Car Brakes. — One of the most important of the components of the motor-car controlling system is usually carried with and forms part of the rear construction, this being the braking means which is utilized to bring the vehicle to a stop when it is desired to arrest forward or backward motion. It will be evident that in a horse-drawn vehicle the animal drawing it can be used as brake, but that in any form of self-propelling conveyance it is essential that some means of stopping be included in the construction. Even if the clutch was operated in such a way that the motor was disconnected from the driving wheels the conveyance would continue to move because it had acquired a certain momentum which would increase in value with the weight of the car and the speed at which it was driven.

There are three brakes provided on the Ford chassis, one of these being a service brake acting on the transmission gear, the other two being emergency members acting on the drums carried by the rear wheels. The service or transmission brake is clearly shown in the view at the top of Fig. 28 in connection with its operating pedal and is also outlined at the bottom of the same illustration which shows the gearing when the top portion of the gear case is removed to expose the three transmission bands. The transmission brake is the one normally used when driving the car and is operated by the right hand pedal of the control assembly which is marked B. When the pedal is pushed forward it constricts an asbestos fabric lined brake band around the drum that also forms the casing for the multiple disk clutch assembly. As this drum is part of the assembly to which the propeller shaft is attached and as this in turn controls the rear wheels through the medium of the bevel pinion carried at its lower end, whenever the transmission assembly is gripped by this brake band it will also retard the movement of the rear wheels and if the brake pedal is pushed tightly enough the friction will be so great that the rear wheels cannot turn and must come to rest even on a steep incline. This is called the "service brake" because it is more generally used than the hand operated or emergency brake acting directly on the rear wheel drums.

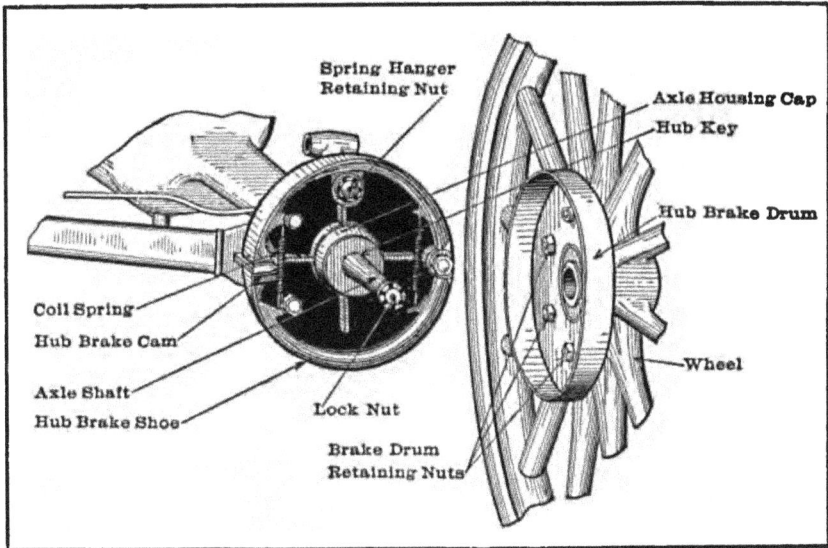

Fig. 34.—End View of Ford Rear Axle with Wheel Removed to Show Emergency Brake Construction.

The emergency brakes are of the type shown at Fig. 34. These consist of a pair of semi-circular cast iron shoe members held together against an anchorage pin and an expanding cam by coil springs as shown. The diameter of the circle formed by these two metal shoes is slightly less than the inside diameter of the brake drum when the brake is not in use. If the hub brake cam is rocked, however, so that instead of lying flat it is moved at such an angle that the brake shoes are spread apart they will grip the internal periphery of the pressed steel brake drum, retarding or entirely stopping the movement of the wheels, depending upon the pressure applied at the end of the band lever and the movement of the actuating cam. As soon as the pressure is released the coil springs bring the brake shoes out of contact with the dnim and permit free rotation of the wheel.

The brake actuating cams are controlled by small levers which are connected with smaller members on the ends of the control shaft which is worked by the hand lever. Rods are utilized to join these levers, these being guided by clips secured to the radius rods. When the hand lever is pulled toward the operator or the rear of the machine it moves the controller shaft and rods forward and pulls the cam operating levers so these spread the brake shoes apart. The emergency brake linkage is interconnected with the clutch actuating pedal so when the handle is placed in a certain position the clutch will be disengaged but the brakes will not be brought into action until a further movement of the hand lever takes place. The handle may be locked in any desired position by a simple ratchet and pawl arrangement at its lower end. This is a good feature as the emergency brakes may be applied to prevent the car from moving when the motor is being cranked or when it is left unattended on a grade. The service brake may be operated at the same time as the emergency brakes are, if desired, though it is only on very steep hills that both brakes can be used to advantage.

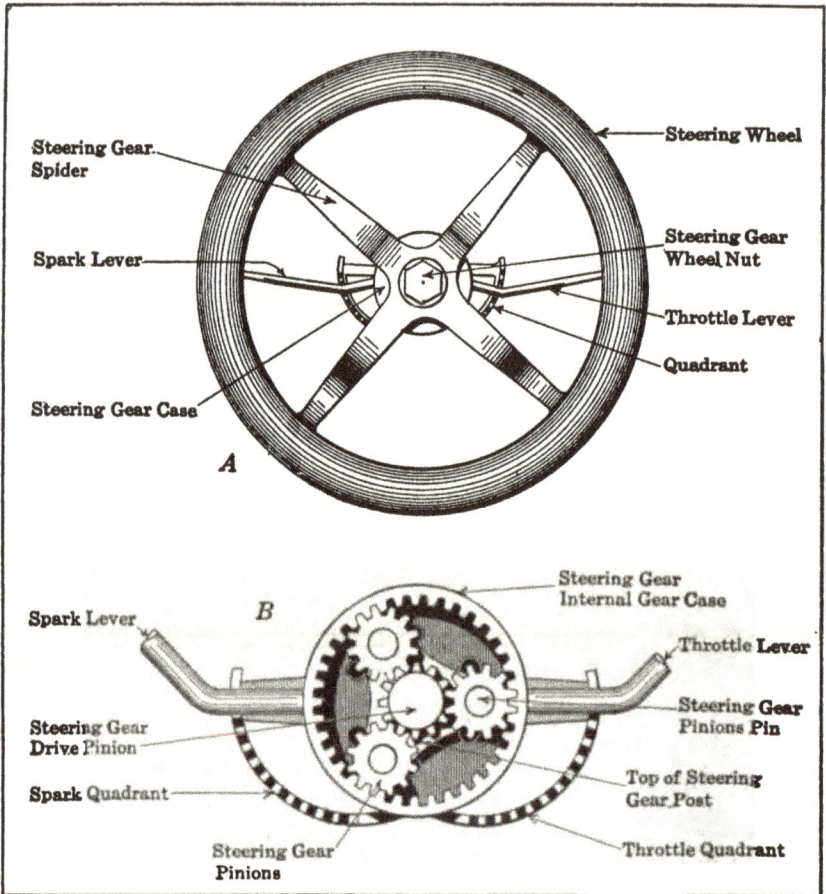

Fig. 35.—Top View of the Ford Steering Gear at A Showing Steering Wheel and Motor Speed Controlling Levers. Planetary Reduction Gearing is Depicted at B which Shows Gear Compartment with Cover Removed.

The Ford Steering Gear. — The manner in which the front wheels are carried by movable steering spindle members has been clearly described in the first chapter, as was the linkage by which the two steering knuckles are

68

caused to move simultaneously when the drag link connecting the front axle with the steering gear is moved by the hand wheel. It was explained that the steering arm at the lower portion of the steering gear was moved by the hand wheel carried at the top of the steering column and shown at Fig. 35, A. The steering post is a metal rod carried inside of the column which is capable of being moved a certain number of degrees to rock the steering arm, this being limited by the travel of the front wheels. The steering column is attached to the dash and is set at such an angle that the hand wheel is brought in a convenient position for the operator. The steering post is housed in a metal tube having an inside diameter sufficiently large to permit of free rotation of the steering post and also to carry the spark and throttle actuating rods which are worked by levers placed below the steering wheel in a position convenient for the operator. The steering wheel consists of a metal spider having four arms which terminate at the oval section rim which is of wood. These arms converge to a boss which forms the center of the steering wheel, a hole being machined in this boss so that the wheel is a tight fit on the member to which it is attached by key and retention nut.

The Ford steering gear differs radically from the conventional forms and is a patented design. The reduction gears which permit of a greater degree of hand wheel movement than of the steering arm are located at the top of the steering column instead of at the bottom as in most other cars. Whereas the worm gear reduction is popular on other types the Ford gearing operates on the epicyclic or planetary principle. The gearing is carried in a compartment immediately under the steering wheel which has a removable cover to permit of examining the gears which require but little attention other than keeping the compartment filled with grease. The construction of this gearing is clearly shown at Fig. 35, B. Four gears of the external spur form are used, one of these being in the center while three surround this member, being carried by studs secured to a triangular plate at the top of the steering post. The casing is provided with teeth on its inner periphery so that it is an internal spur gear. The three outer gears are in mesh with this as well as the central member to which the steering wheel is attached. When the wheel is rotated it turns the center pinion which causes the other three pinions attached to a steering column to rotate also. When the steering wheel is turned to the right the three outer gears are turned in the opposite direction but they are restrained by the internal gear case in such a way that the top of the steering post to which they are fastened moves in the same direction as the hand wheel but travels slower. This provides a certain amount of leverage which makes it easy for the operator to steer the car even under unfavorable road conditions. The spark and throttle levers may be set at any desired position because they work on fixed quadrants attached to the steering column. The spark and throttle control levers do not turn with the steering wheel.

Construction of Ford Tires. — The most common form of tire, and that used on Ford automobiles, is composed of a hollow endless rubber ring or tube of circular section filled with air and protected from wear by means of an outer shoe or casing. The use of air under compression provides a very resilient medium for supporting the vehicle, and of the various forms of rubber tires the pneumatic form is the one that is the most desirable for pleasure cars. The development of the modern automobile may be attributed largely to the advances made in pneumatic-tire construction, as these members made it possible to drive automobiles at high speeds over rough road surfaces without stressing the mechanism unduly or causing discomfort to the passengers. The Ford front tires are 30" x 3", the rear ones 30" x 3½". These are economical as they are easy to buy and give excellent mileage if the car is driven carefully.

The pneumatic tire of the present day is invariably of the double-tube type and is composed of two members, the inner tube and the shoe or carcass. The inner member is utilized to retain the air and is made of a very pure rubber or resilient rubber composition with walls about an eighth of an inch thick for cars of average weight.

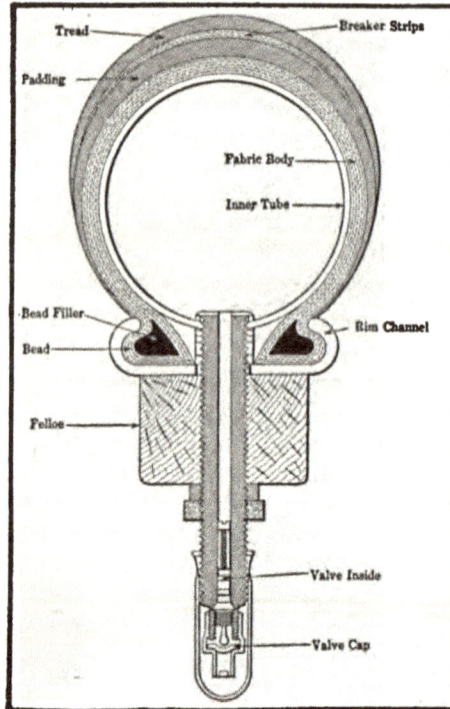

Fig. 36.—Sectional View of Standard Clincher Double Tube Pneumatic Tire Such as Used on Ford Cars.

While this tube is very elastic and is airtight, it would not be strong enough or have adequate resistance to be run directly in contact with the road surface; therefore it is necessary to protect it by a shoe composed of layers of fabric and rubber composition. The shoe member is provided with beads on its inner periphery designed to interlock with the rim channel, as shown on Fig. 36. The main portion of the outer casing is composed of five or more layers of a Sea Island cotton fabric "frictioned" with high-grade rubber composition. This is forced into the mesh of the cloth by machinery so the fabric will be practically waterproof and will join intimately with other plies by a process of vulcanization when the shoe is cured after it is built up. Outside of the fabric body a layer of a very resilient rubber, approximately of crescent form, known as the padding, is provided to give a certain degree of elasticity. On top of this strips of heavy fabric called the "breaker strip" are interposed to offer a certain degree of resistance to any sharp object that might penetrate the tread and go through the padding and into the fabric body if the breaker strips were not interposed to deflect the puncturing object to one side.

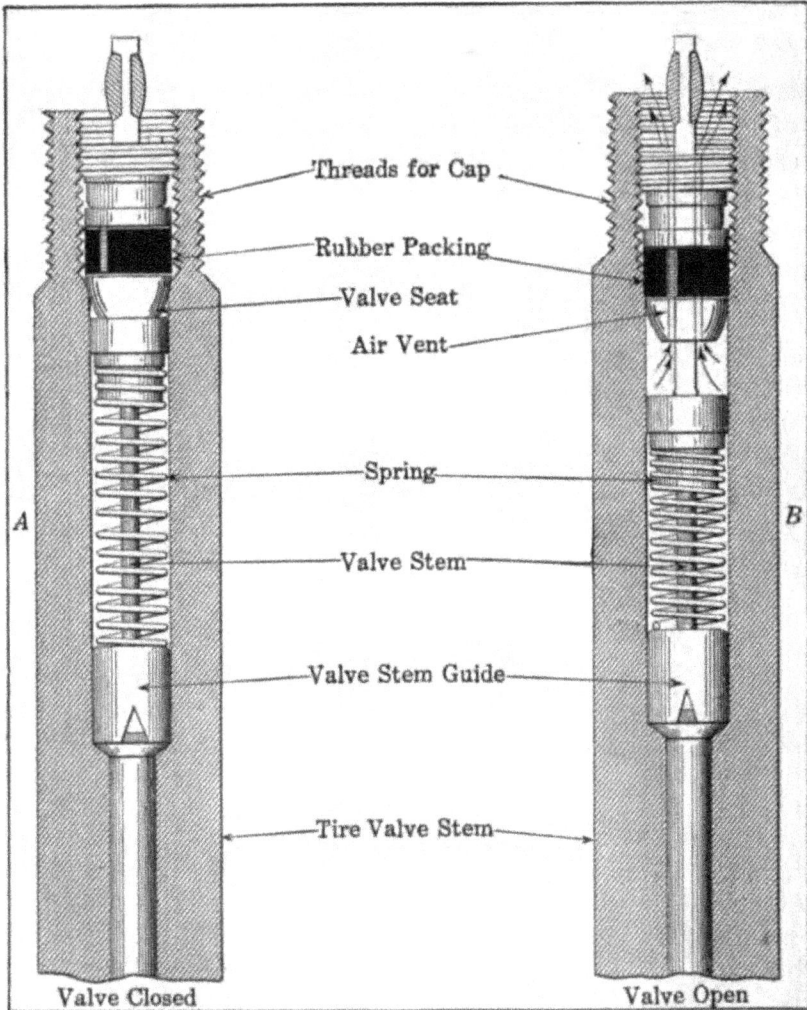

Fig. 37.—Sectional View Showing Construction of Standard Schrader Universal Check Valve For Introducing Air to Pneumatic Tire Inner Tubes. This is Utilized In Practically All Tires of American Manufacture.

The tread is the part of the tire that is subjected to the greatest stress as it must resist the abrading influence of the road and when the tire is used on the rear wheels, the wearing effect of the friction produced by the tractive effort which exists at the point of contact between the driving member and the ground. The tread is of very tough rubber composition and differs from the material used as padding or for the inner tube in that it does not possess a very great degree of elasticity. This quality is sacrificed for that of greater strength and resistance to wear, which is more essential at this point.

The air is introduced in the tire through a simple form of automatic valve which is securely attached to the inner tube. As the inner tube becomes distended by the air pumped into it, it forces the beads of the tire outward and clinches the shoe so firmly in the rim channel that it will be impossible to dislodge it without the use of special tire irons, and then only when the air pressure is relieved from the inner tube. A detailed view of the valve stem is the open and closed positions is shown at Fig. 37 and the construction of this simple fitting can be easily understood. The valve is held against its seat by a tension spring and will only open when the valve stem is depressed by the hand or from the pressure of the air forced against it from a pump when it is desired to inflate the tire. While the air pressure from the pump will be sufficient to force the valve from its seat, the air pressure from the inside of the tire only serves to hold it more firmly in place. Complete instructions for the manipulation, care and repair of Ford tires is given in the next chapter.

Chapter Four - Driving and Maintenance of Ford Cars

There is no point in connection with automobiles that should be covered more completely than that of general operation and maintenance. Even if the motorist does not intend to repair his own car he should be able to care for it intelligently and to locate the various troubles that are apt to materialize while the car is in operation. An engine stop on a lonely road will be a serious matter to one who does not understand the principles of action of the power plant, but only an incident to one who does. If the general principles are understood and the methods of locating ordinary troubles are kept in mind the motorist will be able to keep the car in operation and will fix many of the minor derangements which otherwise would need attention from the repairman, who might not be available when most needed. In presenting this chapter on driving and maintenance the writer wishes to emphasize that endeavor has been made to group practically all the information which the' operator of the Ford ear must have in order to drive intelligently in one chapter. Included with general operating instructions will be found suggestions for systematic location of power plant defects which should be of value to the novice.

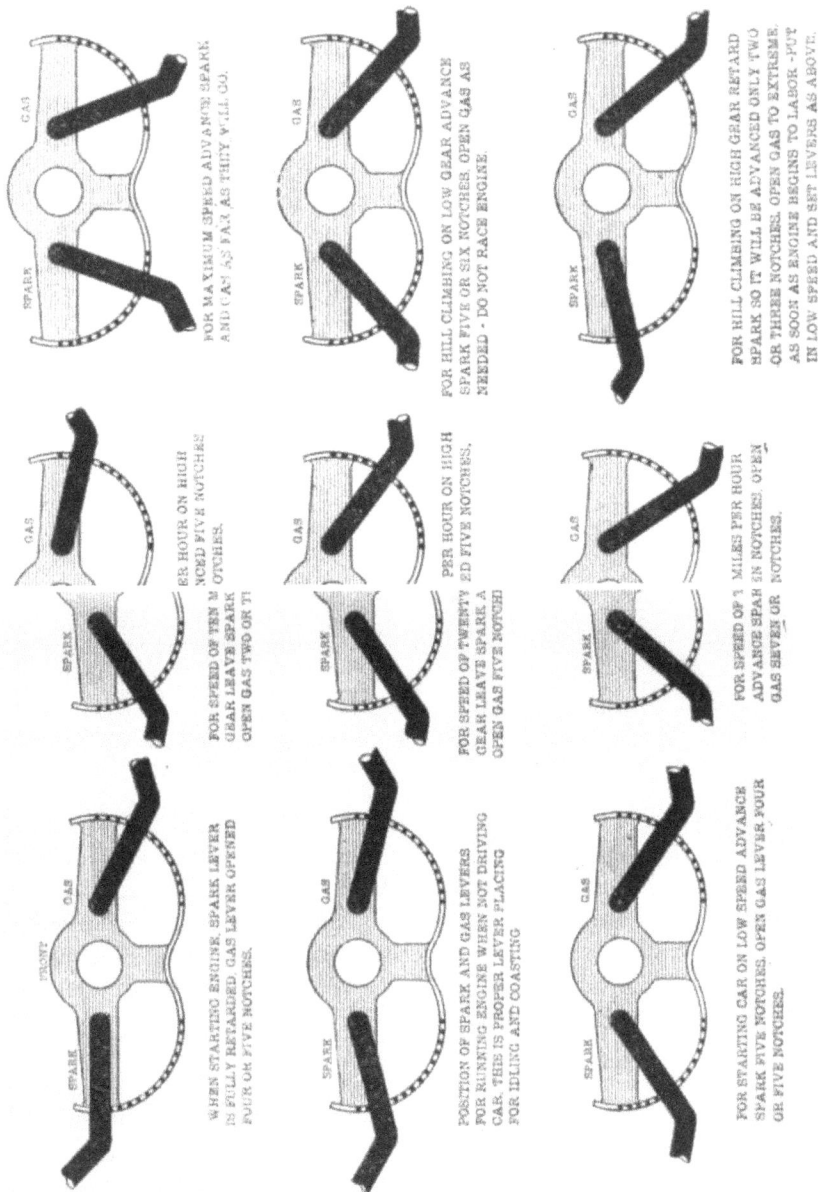

FOR MAXIMUM SPEED ADVANCE SPARK AND GAS AS FAR AS THEY WILL GO.

FOR HILL CLIMBING ON LOW GEAR ADVANCE SPARK FIVE OR SIX NOTCHES OPEN GAS AS NEEDED - DO NOT RACE ENGINE.

FOR HILL CLIMBING ON HIGH GEAR RETARD SPARK SO IT WILL BE ADVANCED ONLY TWO OR THREE NOTCHES. OPEN GAS TO EXTREME AS SOON AS ENGINE BEGINS TO LABOR - PUT IN LOW SPEED AND SET LEVERS AS ABOVE.

...ER HOUR ON HIGH ...NCED FIVE NOTCHES ...OTCHES.

FOR SPEED OF TEN M... GEAR LEAVE SPARK ... OPEN GAS TWO OR T...

...PER HOUR ON HIGH ...ED FIVE NOTCHES.

FOR SPEED OF TWENTY... GEAR LEAVE SPARK A... OPEN GAS FIVE NOTCH...

FOR SPEED OF 1... MILES PER HOUR ADVANCE SPAR... N NOTCHES OPEN GAS SEVEN OR... NOTCHES.

WHEN STARTING ENGINE SPARK LEVER IS FULLY RETARDED GAS LEVER OPENED FOUR OR FIVE NOTCHES.

POSITION OF SPARK AND GAS LEVERS FOR RUNNING ENGINE WHEN NOT DRIVING CAR. THIS IS PROPER LEVER PLACING FOR IDLING AND COASTING.

FOR STARTING CAR ON LOW SPEED ADVANCE SPARK FIVE NOTCHES. OPEN GAS LEVER FOUR OR FIVE NOTCHES.

Fig. 38. Chart Showing Positions of Engine Control Levers on Steering Post Qu₃ for Various Conditions or Car Operation. These are the Average Positions and May Vary Slightly...erent Ford Cars.

Steps Before Starting the Engine. — Before attempting to start the motor there are a number of preliminary precautions to be taken in order to make sure that the car is ready for the road. The gasoline container of the Ford automobile has a capacity of ten gallons, and this should be filled practically to the top, and care should be taken never to start out with less than half a tank full. In order to determine the amount of fuel available a measuring stick may

73

be made according to the dimensions given at Fig. 39, which can be introduced into the top of filler opening to gauge the supply of liquid in the container. The mark to indicate 1 gallon should be made 1 17/32" from the lower end of the stick. The mark to indicate 2 gallons should be made 2 9/16 from the bottom and so on according to the figures given in the diagram. In order to have the liquid show it may be well to paint the stick with a black enamel before marking it. When filling a gasoline tank it is important to use a strainer to prevent water and other foreign substances from reaching the tank interior. Chamois skin is an excellent strainer for gasoline. Three or four layers of fine mesh cotton cloth may also be placed across the month of the funnel if the chamois is not available.

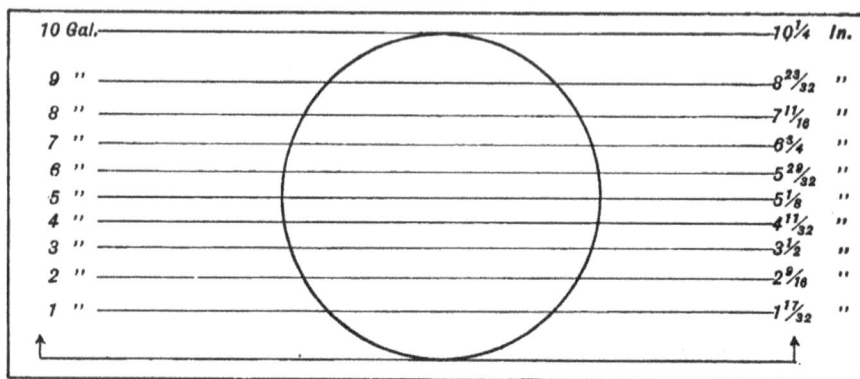

Fig. 39.—Diagram Showing Method of Marking Measuring Stick to Indicate Contents of Ford Ten Gallon Tank.

As gasoline vapor is explosive it is well to make sure that there are no naked flames within several feet of the tank. When filling the tank at night be sure that the oil side lights are extinguished before any fuel is poured. It is important that no matches should be lighted anywhere near where gasoline has been spilled as the air for several feet surrounding the spot has become saturated with highly explosive vapor. The small vent holes in the fuel tank cap should always be free, as if this is plugged up it will prevent the gasoline from flowing into the carburettor, as the fuel would become air-bound in the tank. As has been previously explained, any dirt or foreign matter that would pass into the carburettor will usually fall to the bottom of the sediment bulb on the bottom of the tank, from which it may be drained out by opening a petcock screwed into the bottom of the sediment bulb casting. A shut-off valve is also provided so the gasoline supply in the tank may be conserved at such times as is necessary to remove the feed pipe running to the carburettor. After being sure that the gasoline tank is full the next step is to ascertain that the shut-off valve is opened so that the liquid can flow to the vaporizer.

Before starting out always make sure that a proper supply of medium body, high grade gas engine cylinder oil is poured into the crank case through the breather pipe at the front of the engine. This is clearly shown in

74

the various views of the engine in other chapters and in Fig. 45 herewith. This opening is covered by a brass cap which may be easily withdrawn as it is held in place only by frictional contact. In the back of the lower part of the flywheel casing, which is also the reservoir that holds the oil, are found two petcocks. Be sure the car is on a level floor and pour in oil slowly until it runs out of the upper petcock. Have a can so that the oil running out will be saved. Leave this petcock or small faucet open until the oil stops running, then close it. After the car has been used long enough for the engine to become thoroughly free and easy running, which indicates that all parts have attained the proper bearing, the best results will be obtained by carrying the oil at a level about midway between the two petcocks. If the lower petcock is opened and no oil comes out a proper supply should be immediately placed in the crank case.

Having made sure that there is a proper supply of fuel and lubricating oil the next step is to insure that the cooling supply is adequate. Remove the cap at the top of the radiator and fill with clean, fresh water. If water is taken from a stream or from any other source where there is any question of its cleanliness it is advisable to strain it through muslin or other similar material to keep dirt from getting in and constricting the small bore of the radiator tubes. The Ford cooling system has a capacity for slightly more than three gallons of water. Never run the engine unless the radiator has been filled. The water should be poured in until it runs out of the overflow pipe to the ground which may be accepted as an indication that both radiator and the cylinder water jacket have received an adequate supply. When starting out with a new car or after the engine has been overhauled and parts refitted it is good practice to examine the radiator frequently and to make sure that it is kept properly filled. If the car is used for long distances on slow speed, such as hill climbing and pulling through sand the engine may heat up sufficiently as to boil out the water so that great care should be taken when operating a car under these conditions to replenish the water supply as often as may be necessary. If the only water available contains alkalies and other salts which will deposit sediment in the radiator and water jackets attempt should be made to secure clean soft water such as rain water.

How to Start the Ford Motor. — The essential precautions enumerated having been taken, the first step in starting the motor is to look at the steering wheel and notice the position of the spark and throttle control levers which are clearly shown in inset, Fig. 42. The right hand lever is called a "throttle," as it controls the amount of gaseous mixture drawn into the motor. When the power plant is in operation the nearer the operator this lever is the faster the engine will turn and the greater the power output. The left hand lever controls the spark which should be in retard position or at its extreme position away from the operator when starting the motor. It is possible in many cases to advance this lever three or four notches by moving it toward the seat without any danger of injury in cranking. The throttle lever should be placed about four or five notches down to secure easy starting. The reason it is desirable not to advance the spark control lever too far is that the engine may kick back. This may result in damage to the wrist or arm of the person cranking the motor unless care is taken in the method of handling the starting crank. Before cranking the engine one should make sure that the emergency brake lever is pulled back as far as it will go. When in this position the clutch is out and the hub brakes are engaged, which prevents the ear from moving. After inserting the switch key in the switch on the coil box, throw the switch lever as far to the left as it will go toward the point marked "magneto." The engine cannot be started until the electrical circuit is complete. If batteries are used for ignition as an auxiliary it may be possible to start the engine easier on the battery current, though a very easy start may be secured on the magneto provided the coil vibrators are properly adjusted.

Fig. 40.—Illustrating Correct Method of Grasping Starting Crank to Avoid Injury Due to Back Kick.

After having put the switch in either battery or magneto position the next step is to crank the engine by lifting on the starting crank at the front of the car. Take hold of the handle and push the crank in toward the car until you feel the ratchet on the crank engage with the pin passing through the crank shaft. The crank handle should be pulled upward with a quick swing. The proper method of grasping a crank is shown at Fig. 40. It will be observed that the crank is grasped in the left hand and that any tendency to backfire will pull the handle out of the hand by opening the fingers. The improper method of cranking is shown at Fig. 41. In this case the right hand is placed around the crank handle and the engine is started by pushing down against the compression instead of lifting up on the handle to overcome the compression resistance. It will be evident that if the spark advance lever is set so that an early explosion obtains this will drive the handle vigorously backward, which force is partially resisted by the tensed arm of the operator. There are times when it is necessary to turn the crank vigorously which is called "spinning" the engine. In this case be sure that the spark lever is fully retarded, otherwise a sudden backfire may cause injury.

Fig. 41.—Showing Wrong Method of Exerting Pressure on Crank When Starting Motor.

If the engine has been standing for a time it is advisable to prime the carburettor by pulling on the small wire at the lower left corner of the radiator while giving the engine two- or three-quarter turns with the starting handle. In this case the crank should be grasped by the right hand but care should be taken to only pull up against the compression. In cold weather gasoline does not evaporate very readily so it is somewhat more difficult to start a motor under these conditions. The method recommended by the Ford company for starting the engine when cold is to turn the carburettor dash adjustment one quarter turn to the left in order to allow a richer mixture of gasoline to be drawn into the cylinders, then to hold out the priming rod which projects through the radiator while the crank is whirled vigorously. Another method is as follows: Before throwing on the magneto switch close the throttle lever, hold out priming rod while you give crank several quick turns, then let go of priming rod, place the spark lever in third or fourth notch, advance throttle lever several notches, throw on the switch and crank briskly.

After starting the motor it is advisable to advance the spark half way down the quadrant and to let the motor run until thoroughly heated up. If one starts out with a cold motor it is not likely to have much power and it would be easy to "stall" it. It is said that the advantage of turning on the switch last after priming is that there is plenty of gas in the cylinders to keep the motor running. After the motor is warmed up the carburettor adjustment should be turned back to the proper running position. If, for any reason, the engine is warm and does not start readily it is probably because the engine has been flooded with an over rich gas mixture. The remedy for this condition is to turn the carburettor adjusting needle down by screwing the needle valve on the dash to the right until the needle seats in the carburettor. Crank the engine briskly to exhaust the rich gas, then throw on the switch and start the engine. As soon as the cylinders fire turn back the needle to the normal running position.

If the engine fails to start the following defective conditions may be responsible: Water in the gasoline; water or hardened oil in commutator; coil vibrators out of proper adjustment; gas mixture too thin; gas mixture too rich; magneto contact point in transmission cover raised because of foreign matter or short circuiting by a piece of wire from brake lining; gasoline supply shut off; water frozen in gasoline tank sediment bulb; poor contact at coil switch; loose magneto wire leading to coil; loose timer wires; engine too cold to properly vaporize gas (only in zero weather). Should the engine start, run for a time and then stop suddenly, one should make sure that there is plenty of fuel in the gasoline tank. The trouble may be a flooded carburettor; dirt in carburettor or feed pipe; magneto wire loose at either terminal; magneto current collecting point obstructed; engine overheated on account of insufficient oil or water supply.

If the engine lacks power and runs irregularly, which is called "skipping" at low speed it may be due to: imperfect gas mixture; dirty spark plugs; poorly adjusted coil vibrators; poor compression; air leak through intake manifold; weak exhaust valve springs, too little clearance between valve stem and operating push rod; spark plug points too near together.

If the engine misfires at high speed, it may result from imperfect contact in the interior of the commutator; too much air gap between the points of the spark plugs; imperfect gas mixture or poorly adjusted vibrators. When an engine overheats, the most common condition is running with too rich gas mixture and retarded spark. Other troubles are: insufficient lubricating oil;

not enough water in the radiator; fan belt too loose or slipping; water circulation poor, owing to sediment in radiator tubes; or carbon deposits in combustion chambers. These carbon deposits may be also present on the piston head and will result in loss of power as well as produce knocking sounds. If a loud knock is evident it is usually due to a loose connecting rod or crank shaft bearing or running with the spark advanced too far and it is always the sign of a badly overheated engine.

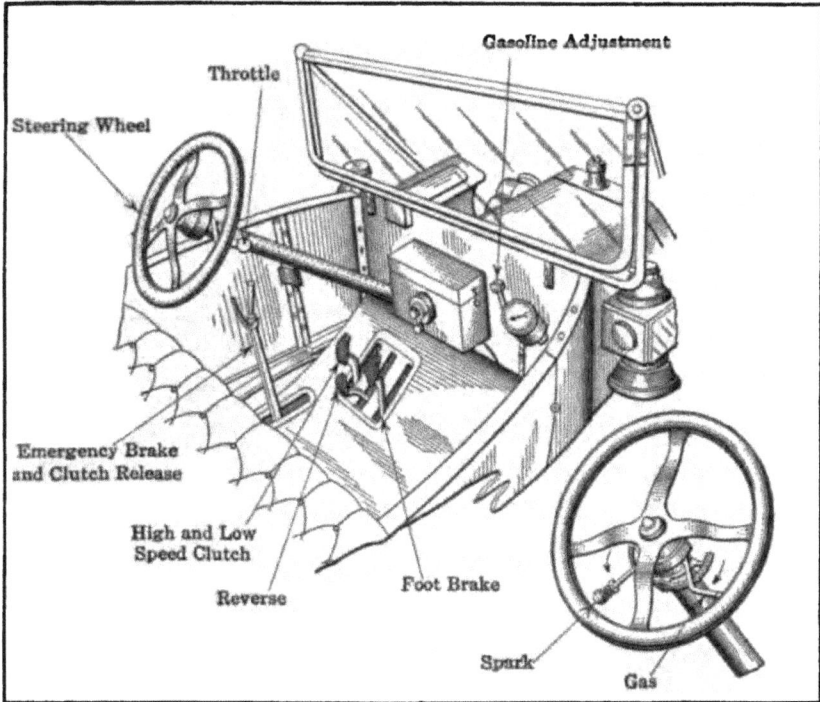

Fig. 42.—The Control System of the Ford Model T Car.

Controlling the Ford Car. — The Ford car is one of the most popular of moderate-priced automobiles and over 600,000 of the Model "T" are now on the road. The control system of this car is extremely simple and yet it is different from that of any other automobile. The gearset, which has been previously described, is a planetary type which gives two forward speeds and a reverse motion. The conventional form of steering wheel is used to control the direction of ear travel, and spark and throttle levers are mounted on steering column beneath the wheel to control the speed of the power plant. It is in the method of obtaining the various speed ratios that the control system is distinctive. As will be seen by referring to Fig. 42, three pedals and a hand lever are provided on the left side of the car. The pedal on the extreme left side controls the high and low-speed clutches and is marked "C." That nest to it, which is marked "R," is used to constrict the reverse band of the transmis-

79

sion and obtain reverse motion. The pedal at the right, which is provided with a letter "B" cast on its surface, is used to apply the foot brake.

The hand lever engages the high speed or direct drive clutch when thrown forward and when pulled back it actuates the emergency brake which cannot be applied without releasing the direct drive clutch. The lever may "Be set in a neutral position and the clutch will be released without applying the brake when it is approximately vertical. When the high speed is in and the hand lever is thrown way forward the high-speed clutch may be released by a light pressure on pedal "C" and a further movement of this pedal will apply the low speed. Thus one pedal gives control of both high and low speeds forward and the clutch can be released in exactly the same manner as that of a sliding gear driven car when it is desired to slow up such as for turning a corner, ascending a hill or passing another vehicle.

Before starting the car the hand lever must be in a vertical position, this releasing the clutch and applying the emergency brakes. To start the car, after the engine has been started, the foot is placed on the clutch pedal to keep it in a neutral position while the hand lever is thrown as far forward as it will go. The engine is then accelerated and the clutch pedal is pushed forward until the slow speed band tightens around the drum of the transmission and the car gathers headway on the lower ratio. After it has attained a certain momentum the clutch pedal is allowed to drop back gradually into the high-speed position. The foot may then be removed until such times that the clutch must be disconnected. Before applying the foot brake, which is done by pressing with the right foot upon the pedal marked "B" the clutch pedal should be put in neutral position with the left foot.

To reverse the car, it must first be brought to a standstill. The engine is kept running and the clutch is disengaged with the hand lever, which is placed in the neutral position but not pulled far enough back to apply the emergency brake. The reverse pedal marked "R" is then pushed forward with the left foot leaving the right one free to use on the brake pedal if needed. To stop the car, the throttle is closed so the engine will not race; the high speed is released by pressing the clutch pedal forward into its neutral position and applying the foot brake slowly, but firmly, until the forward motion of the car is arrested. It is imperative that the foot be retained on the clutch pedal until the hand lever is pulled back to its neutral position. The placing of the spark and throttle levers is clearly shown in the inset in the right-hand comer of the cut, both levers being pulled back to accelerate the motor and pushed forward to slow it down.

General Driving Instructions. — The following instructions apply to all types of gasoline automobiles as much as to the Ford and may be followed to advantage by all motorists. The gear-shift lever should always be placed in a neutral position when the car is stopped, whether it is left alone or attended. The clutch should always be applied gradually and as slowly as possible because too sudden or harsh engagement will produce stresses that will injure the tires or driving mechanism of the chassis. Never allow the engine to race or run excessively fast when changing speeds, and it is well not to undertake to change speeds with either motor or car running at high speed. When changing down, i.e., from a higher to a lower gear, allow the car to slow down until its speed is about the same as that which will be produced by the lower gear ratio desired before the clutch is engaged to produce the lower speed.

Always drive a car slowly and cautiously until you are thoroughly familiar with the control mechanism and the methods of stopping the car. When driving up grades on the high speed, if the motor shows any tendency to labor, shift back into the lower gear ratio which has been provided for that purpose. Many motorists believe that the best test of a car's ability is to rush all hills or bad spots in the roads on the direct drive. It should be remembered, that the lower speed ratio was provided for use at all times when employing the high speed might produce injurious stresses in the motor. All unusual noises should be investigated at once as these sounds usually presage more or less serious trouble. A gasoline car should never be driven with a slipping clutch, and it is imperative that the brakes and steering gear be frequently inspected to make sure that they are in proper order.

One should never attempt to drive Ford cars at high speeds unless the tire casings are in perfect condition and the road surfaces good. When driving on clay or muddy roads, or on wet asphalt, care must be taken in turning corners and the car should be driven cautiously to avoid dangerous side slipping or skidding. When driving on unfavorable highway surfaces always keep one side of the car on firm ground, if possible.

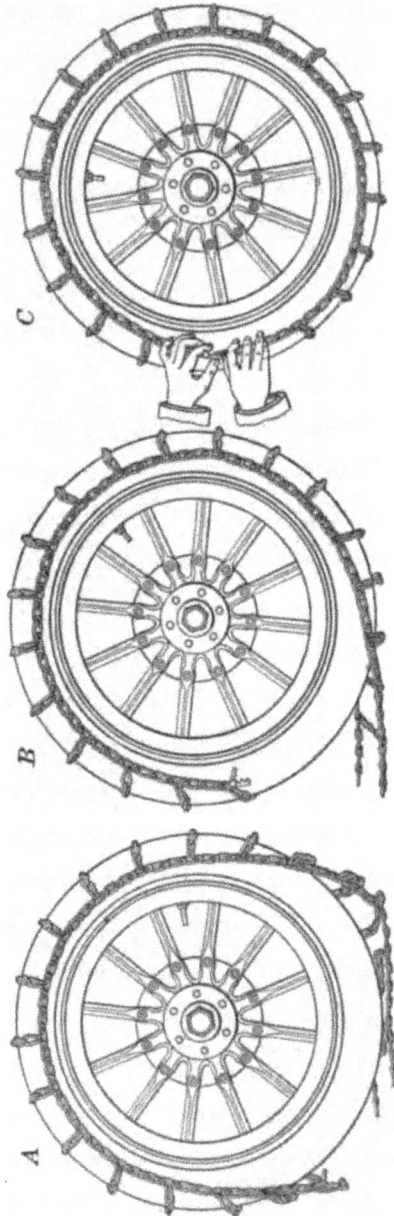

Fig. 43.—Showing Method of Applying Non-Skid Chain to Driving Wheels to Insure Adequate Traction on Wet or Slippery Roads.

Brakes should always be carefully applied, especially if the road surfaces are wet. An automobile should never be brought to a stop in mud, clay or sand, snow or slush, if it can be avoided. Whenever road conditions are unfavorable the smooth tread tires of the driving wheels should always be fitted with chain-tire grips, as shown at Fig. 43, to insure having adequate traction.

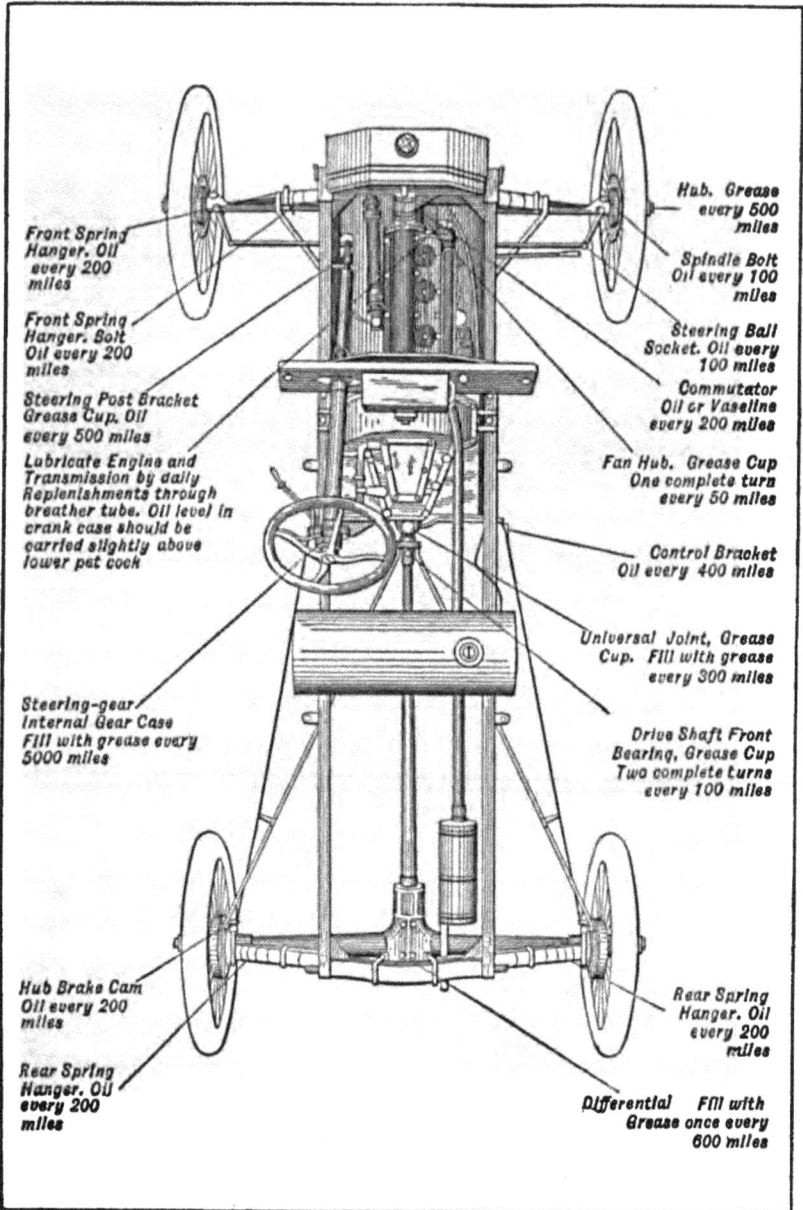

Hub. Grease every 500 miles

Spindle Bolt Oil every 100 miles

Steering Ball Socket. Oil every 100 miles

Commutator Oil or Vaseline every 200 miles

Fan Hub. Grease Cup One complete turn every 50 miles

Control Bracket Oil every 400 miles

Universal Joint, Grease Cup. Fill with grease every 300 miles

Drive Shaft Front Bearing, Grease Cup Two complete turns every 100 miles

Rear Spring Hanger. Oil every 200 miles

Differential Fill with Grease once every 600 miles

Front Spring Hanger. Oil every 200 miles

Front Spring Hanger. Bolt Oil every 200 miles

Steering Post Bracket Grease Cup. Oil every 500 miles

Lubricate Engine and Transmission by daily Replenishments through breather tube. Oil level in crank case should be carried slightly above lower pet cock

Steering-gear Internal Gear Case Fill with grease every 5000 miles

Hub Brake Cam Oil every 200 miles

Rear Spring Hanger. Oil every 200 miles

Fig. 44.—Plan View of Ford Model T Chassis Showing Important Points Requiring Lubrication and When This Attention is Needed.

All motorists should familiarize themselves as much as possible with the mechanism of their cars and should feel competent enough to make the ordinary adjustments and minor repairs before any long trips are attempted. A full equipment of tools and spare tires and casings should be carried at all times. It is well to remember that the manufacturer of the car has issued a set of instructions for its care and maintenance, and these should be followed as closely as possible because intelligent care of any piece of machinery means long life and reliable service and the automobile is no exception to the rule.

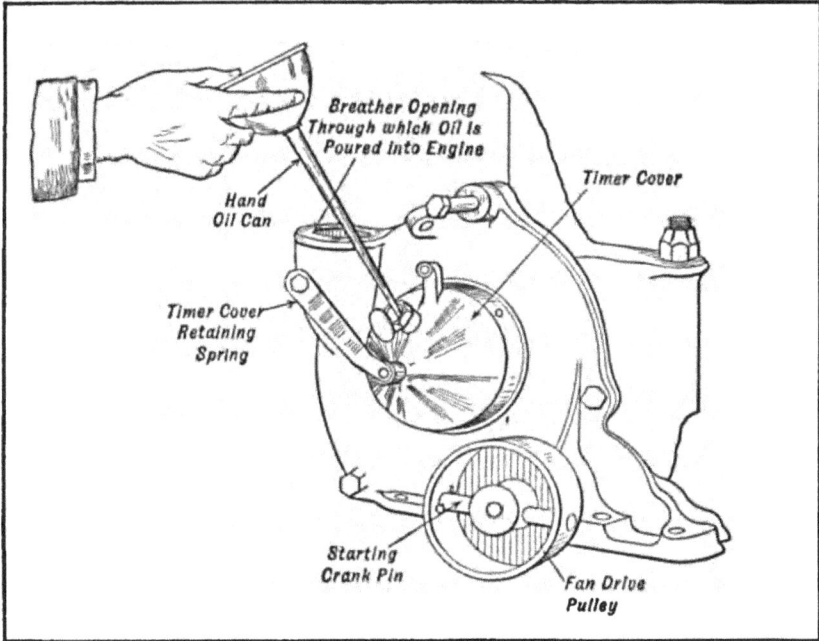

Fig. 45.—Method of Oiling the Ford Commutator or Timer With Light Oil. Note Breather Opening Back of Timer Through Which Oil is Poured Into Crank Case.

Suggestions for Oiling. — One of the most important points to be observed in connection with gasoline automobile operation is that all parts be oiled regularly. It is not enough to apply lubricant indiscriminately to the various chassis parts, but it must be done systematically and logically to secure the best results and insure economical use of lubricant. The most important parts are the power plant and transmission system and the engine is but one point in the car that must be properly oiled at all times to obtain satisfactory results. Some of the running-gear parts are relatively unimportant, others demand regular inspection and oiling. A very comprehensive oiling chart is presented at Fig. 44, this showing practically all of the points that require oil as well as giving instructions regarding the character of the lubricant needed and how often it should be applied. Some of the points are governed by special instructions, these being the transmission case, timer, and rear axle. Use only the best medium body cylinder oil in the Ford motor. The writer has obtained excellent results by putting in a quart of lubricant to every five gallons of gasoline, the oil was introduced through the breather pipe every time that amount of fuel was placed in the tank.

Neither the transmission case nor the differential ease on the rear axle should be filled with heavy "Dope" widely sold, which may contain fiber or cork particles to make for more silent operation. If gears are noisy it is either because they are worn or out of adjustment, and the use of the nostrums and freak lubricants will not improve their operation. The rear axle differential housing should be filled with as light mineral grease as it is possible to get, those having about the consistency of vaseline and containing graphite being most desirable as lubricants. Light oils should not be used in the rear axle housing, because these will not stay in place but leak out over the brakes and will not have sufficient body to cushion the gear teeth. The only other point on the chart which needs explanation is lubrication of the timer interior. This should be oiled, as it is a roller contact form and a few drops of magneto or 3-in-l oil should be applied to the roll and the contact segments once a week. The timer case should be removed and all old, dirty oil washed out with gasoline before fresh lubricant is supplied. Never use graphite grease or any heavy oil in a timer case because these will not only interfere with regular ignition by short circuiting the current, but they will clog up the timer and prevent the roller establishing proper contact with the segments. After a car is oiled it is well to go over all the exposed joints with a piece of cloth to remove the accumulation of surplus oil on the outside of the parts which serves no useful purpose and which only acts to attract and retain dust and grit. The instructions given on the chart can be followed to advantage on many types of gasoline cars, though, of course, the different constructions will have to be treated as the peculiarities of design dictate.

A systematic method of lubricating the various parts is much to be preferred to the usual haphazard way in which the cars are oiled. The speedometer may be used to check off the mileage made by the car and if a note is made of the distance covered no trouble should be experienced in following the chart at Fig. 44. The simplicity of the Ford car makes for easy lubrication as the entire mechanism can be thoroughly oiled in less than five minutes. The places needing lubrication, itemized under the heads of mileage covered, follow:

Lubricate After 200 Miles' Driving.

Lubricant	Number	Name of Parts
Oil	2	Front axle, steering knuckle pivots or spindle bolts.
Oil	2	Front spring shackles and bolts.
Oil	2	Yokes of tie rod.
Oil	1	Steering ball socket.
Oil	1	Commutator or timer.
Oil	2	Rear hub brake cams.
Oil	2	Rear spring shackles and bolts.
Turn grease cup	1	Fan hub.

Lubricate After 500 Miles' Driving.

Turn grease cup	1	Steering post bracket.
Turn grease cup	1	Universal joint of shaft.
Turn grease cup	1	Driving shaft front bearing.
Grease	2	Front wheel hubs.

Lubricate After 1,000 Miles' Driving.

Grease	1	Differential housing.

Lubricate After 2,000 Miles' Driving.

Oil	1	Control bracket.

Lubricate After 5,000 Miles' Driving.

Grease	1	Steering gear internal gear case.

Lubricate Daily.

Oil	1	Motor.

Lubricate Occasionally.

Oil	1	Fan belt shaft.
Oil	1	Fan belt pulley.
Oil	1	Crank handle bearing.
Oil	4	Yokes of brake rods.

In referring to the process of oiling, this means using a sufficient quantity to lubricate the bearing parts thoroughly, and turning the grease cups means that these assumedly contain grease. Greasing means packing the bearing or housing until it is filled. The caretaker can examine the condition of the grease cups, and when these are found properly filled they should be screwed down.

Filling and screwing down each cup three times, to insure ample supply of lubricant being in the bearing, is a safe practice to follow. The fourth time filled the cup can be left for the stated mileage interval. Care should be taken to wipe the cups clean before filling, to prevent dirt being carried into the bearings, and the oilers should be cleaned with equally good reason.

85

The best attention can be given at the end of the day's or night's driving, which will require but very little time, for conditions will all be favorable. The engine should be wiped clean while it is warm, for the oil or grease and dust accumulated will be soft enough to remove easily. The oil cups should be filled, all the grease cups turned according to the mileage for the day, the fuel supply renewed, the water supply in radiator replenished, and the oil in the engine case brought to the required level. The next time the machine is wanted it will be ready for use and the owner will know that it can be driven 200 miles or more with absolute certainty that it will have sufficient oil at all points except the engine which should be looked at at the end of 100 miles to make sure there is enough oil in the flywheel compartment.

Systematic attention to oiling and greasing, such as has been described, will so familiarize a man with the normal conditions that he will note whether or not there is wear of any of the moving parts, and one will find that there is usually need of tightening nuts and screws that will slacken, no matter how well they have been set, and these ought to be tightened. It 'is easy to discover loose parts while oiling and take immediate steps to remedy the defective condition.

Winter Care of Automobiles. — While motoring throughout the entire year is not unusual, many owners of cars, especially in those portions of the country where the winter climate is exceptionally severe, put up their car for the winter period. If the car is kept in service the most important thing to do is to provide some good anti-freezing compound in order to prevent the water in the radiator and cylinders from congealing. There is some difference of opinion regarding the best solution to use to prevent cracked water jackets and burst radiators. Before we attempt to answer the questions often asked regarding the best anti-freezing compound, it will be well to consider the requirements of such compounds. To begin with it should have no deleterious effects on the metals or rubber used in the circulating system. It must be easily dissolved or combined with water, should be reasonably cheap, and not subject to waste by evaporation or be of such character that it will deposit foreign matter in the pipes. The boiling point should be higher than that of water to prevent boiling away of the solution at comparatively low temperature.

Solutions of calcium chloride have been very popular with motorists, and the writer will first discuss the use of this substance. The freezing point of the solution depends upon the proportion of the salt to the water. An important factor to be considered is that if the parts of the circulation system are composed of different metals there is liable to be a certain electrolytic action between the salt and the dissimilar metals at the points of juncture, a certain corrosion taking place, and the intensity of this corrosive effect is dependent upon the strength of the solution. As calcium chloride is derived from hydrochloric acid, which has very strong effect on metals, and as there may be particles of the free acid in the solution, a certain undesirable corrosive action may take place. In using calcium chloride when compounding an anti-freezing solution care must be taken that commercially pure salt is employed, as the cruder grades will liberate a larger percentage of free acid. The mistake should not be made of using chloride of lime, which has much the same appearance, but the corrosive action of which is very great.

It is well to test a solution of calcium chloride for acid before placing it in the radiator. A piece of blue litmus paper may be obtained at any drug store and immersed in the solution. If the paper turns red it is a sign that there is acid present. Acid may be neutralized by the addition of a small quantity of slacked lime.

The solutions may be made in these proportions:

Two pounds of salt to the gallon of water will freeze at eighteen degrees Fahrenheit.

Three pounds of salt to the gallon of water will freeze at one and five-tenths degrees Fahrenheit.

Four pounds of salt to the gallon will freeze at seventeen degrees Fahrenheit below zero.

Five pounds of salt to the gallon will freeze at thirty-nine degrees Fahrenheit below zero.

It must be remembered that the more salt to the solution, the greater the electrolytic effect and the greater the liability of the deposit of salt crystals which may obstruct the free flow of the liquid.

Glycerine is usually considered quite favorably, but it has disadvantages. It often contains free acid, though the action on metals will be imperceptible in average solutions. While it does not attack metal piping to any extent it is sure destruction to rubber hose and should not be used in a car in which part of the circulation system piping is of rubber. Glycerine is expensive and it is liable to decompose under the influence of heat and proportions added to the water must be higher than of some other substances.

Denatured alcohol is without doubt the best substance to use as it does not have any destructive action on the metals or rubber hose, will not form deposits of foreign matter, and has no electrolytic effect. A solution of sixty per cent, water and forty per cent, alcohol will stand twenty-five degrees below zero without freezing. The chief disadvantage to its use is that it evaporates more rapidly than water and the solution is liable to become too light as proportions of alcohol to water is concerned. The percentages required are shown in the following sentences:

Water, ninety-five per cent.; alcohol, five per cent, freezes at twenty-five degrees Fahrenheit.

Water, eighty-five per cent.; alcohol, fifteen per cent, freezes at eleven degrees Fahrenheit.

Water, eighty per cent.; alcohol twenty per cent, freezes at five degrees Fahrenheit.

Water, seventy per cent.; alcohol, thirty per cent, freezes at nine degrees Fahrenheit below zero.

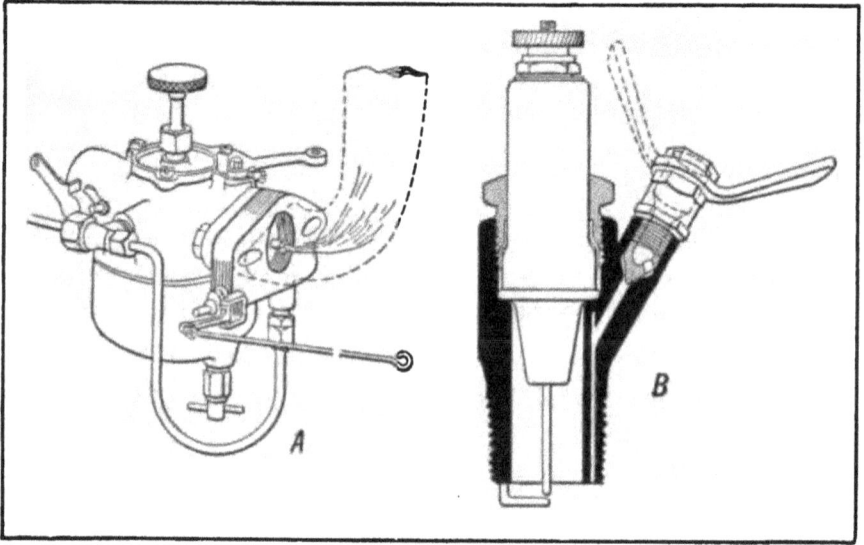

Fig. 46.—Devices to Facilitate Starting Ford Motor in Cold Weather. A.—Injex Primer. B.—Spark Plug With Priming Valve Attachment.

Water, sixty-five per cent.; alcohol, thirty-five per cent.; freezes at sixteen degrees Fahrenheit below zero.

Various mixtures have been tried of alcohol, glycerine and water, and good results obtained. The addition of glycerine to a water-alcohol solution reduces liability of evaporation to a large extent, and when glycerine is used in such proportions it is not liable to damage the rubber hose. The proportions recommended are a solution of half glycerine and half alcohol to water. The glycerine in such a solution will remain practically the same, not being subject to evaporation, and water and alcohol must be supplied if amount of solution in radiator is not enough.

The freezing temperatures of such solutions of varying proportions are as follows:

"Water, eighty -five per cent.; alcohol and glycerine, fifteen per cent.; freezes at twenty degrees Fahrenheit.

Water, seventy-five per cent.; alcohol and glycerine, twenty-five per cent.; freezes at eight degrees Fahrenheit.

Water, seventy per cent.; alcohol and glycerine, thirty per cent.; freezes at five degrees Fahrenheit below zero.

Water, sixty per cent.; alcohol and glycerine, forty per cent.; freezes at twenty-three degrees Fahrenheit below zero.

The proper proportions to he used must of course he governed by conditions of locality, but it is better to be safe than sorry, and make the solutions strong enough for any extreme that may be expected.

After due care has been taken with the cooling system to prevent freezing, the next point to observe is the lubrication of the motor. This will depend on the grades of oil which are normally employed. As a general rule it is well to use a lighter grade in winter than that utilized during warmer weather. If an acetylene lighting system utilizing a gas generator is fitted it is necessary that the water used in the water tank or the water jacket provided on some generators be drained off and replaced with a solution of denatured alcohol and water of the proper consistency for the degree of temperature liable to be met with.

During cold weather, a certain amount of difficulty is always experienced in starting the car, especially when one considers the low grade of gasoline used at the present time. The Ford engine is not provided with petcocks through which gasoline may be injected as in other automobiles. Special spark plugs may be obtained having a petcock incorporated with the plug body or a special primer may be placed between the carburettor and manifold, as shown at Fig. 46. Pulling a wire when cranking a car equipped with the primer permits gasoline to flow directly to the intake manifold as shown. In extreme cold weather many motorists disconnect the fan belt in order that the air draught through the radiator will not cool the water to such a point that the engine will not run efficiently. Other motorists provide some form of a lined leather shield for the front of the radiator.

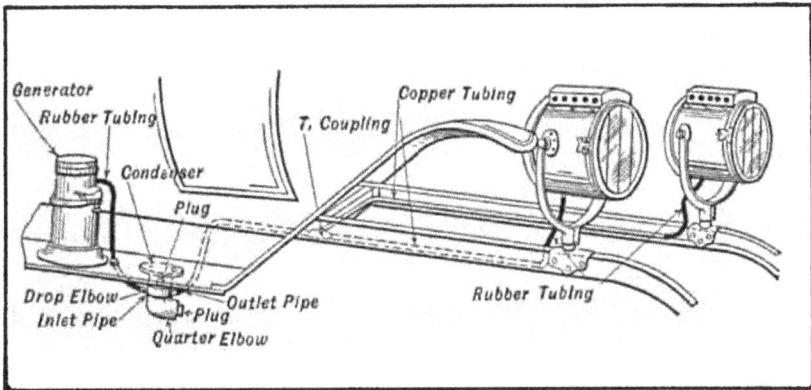

Fig. 47.—Acetylene Gas Lighting System Similar to That Used for Ford Lights on 1910 to 1914 Models. 1915 Ford Cars Have Electric Head Lights.

The Ford Lighting System. — The system of lighting supplied with the Ford car includes 3 oil lamps, two at the dash and one at the rear. The headlights of models made previous to 1915 are of the acetylene gas burning type deriving the gas from action of water on calcium carbide in a simple generator carried on the running boards. Pure calcium carbide will produce about 5.5 cubic feet of gas per pound of carbide decomposed, but the commercial product seldom yields more than 4.5 cubic feet. Acetylene is a very brilliant illuminating gas and gives a white light of about 240 candlepower if burned at the rate of five cubic feet an hour. The strength of illumination can be better judged by comparing it with that produced by burning five cubic feet of good coal gas in the same period of time which will result in only 16 candle-power. A special form of burner is used in the Ford automobile headlights, which mixes a certain amount of air with gas and the brilliant white light produced is intensified and projected by means of a lens mirror placed at the back part of the lamp. This lens serves to collect and concentrate the rays of light from the flame into a beam composed of parallel rays which have great illuminating power, and which will light up the road for several hundred feet ahead of the car and permit higher speeds with safety than would be possible with the feeble glimmer of oil lamps.

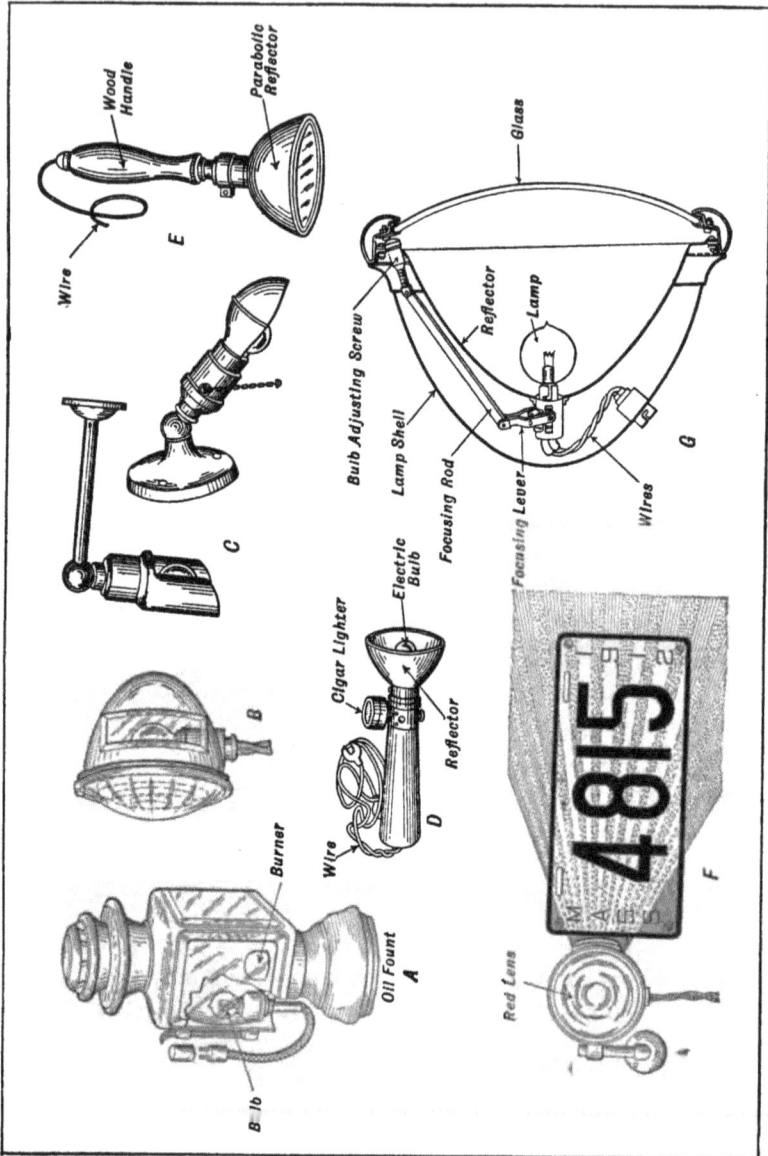

Fig. 48.—Lamps and Fixtures Adapted for Electric Current.

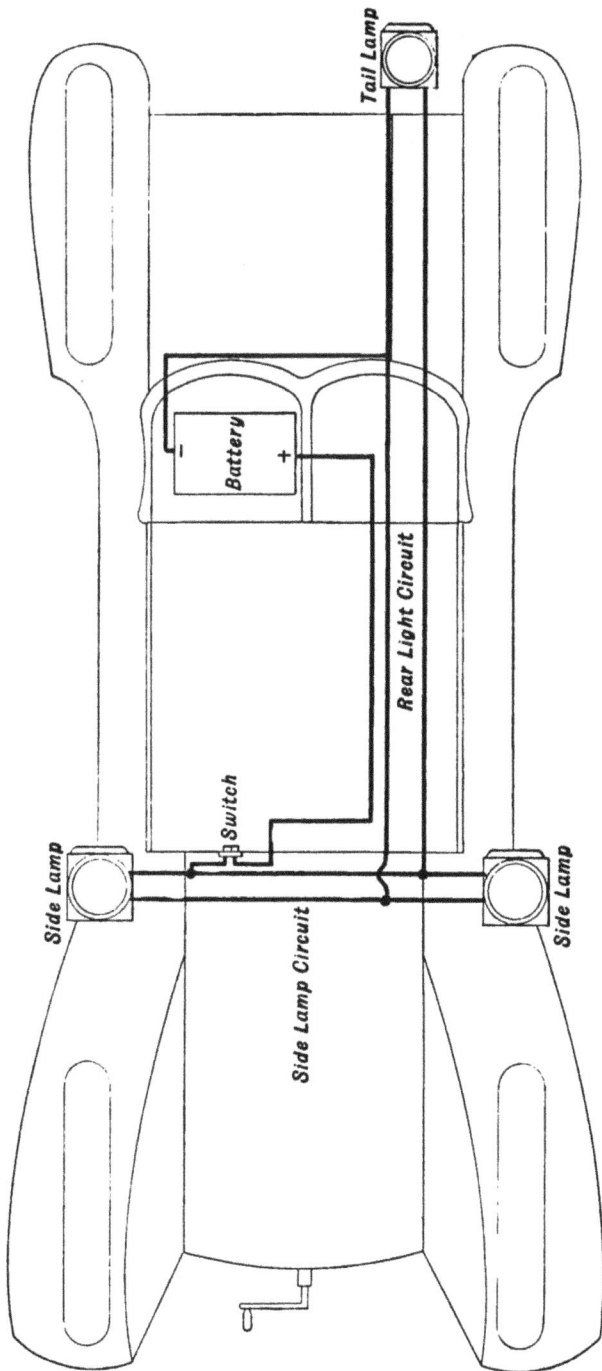

Fig. 49.—Simple Wiring Diagram Showing Method of Installing Storage Battery or Multiple-Series Dry Battery for Operating Electric Side and Tail Lamps.

The generator employed and its mode of action may be easily understood. It consists of a water tank and separate compartments for carbide and as soon as the two come in contact the chemical begins to decompose and acetylene gas is liberated while lime dust collects in the bottom of the generator as a residue. The gas collects in a reservoir and forces its way through a filter chamber filled with wool or similar material which filters the gas. The gas is also cooled before it reaches the lamps because the gas outlet pipe and filter is surrounded with water. When the shut-off valve is opened it permits the water which is carried in the upper chamber to drop into the carbide basket through a perforated tube. If the pressure in the intermediate compartment is normal atmospheric pressure, the water will drop freely onto the carbide until considerable gas is liberated. The generator will continue to supply gas as long as the supply of water and carbide lasts. The jarring produced by car movement sifts the lime to the bottom, and tends to keep the carbide crystals clean so they may be readily acted upon by water. The generator must be cleaned after every trip in which it is used and all lime dust removed and carbide remaining freed of dust. The best method of piping is shown at Fig. 47, the water separator being a fitting that must be furnished by the owner as it is not supplied with the car. This keeps water out of the pipe line and prevents lamps from flickering.

Electric Lighting for Ford Cars. — Many owners of Ford cars have fitted electric lights instead of the kerosene lamps and gas lights regularly furnished up to this year. A number of attachments have been offered designed to fit the gas head lights; these consist of parabolic reflectors and electric bulbs intended to be run from the same magneto that furnishes ignition current. Previous to the year 1915, the Ford Motor Company did not recommend the use of the magneto current for electric lighting inasmuch as it was stated that this interfered with ignition. The 1915 Ford cars, however, are equipped with electric head lamps as a regular fitting, the current being derived from the Ford magneto, which has been redesigned with a view to using larger magnets and thus producing more electrical energy than is needed for ignition purposes. As electrical lighting is general on all makes of cars and has so many advantages, many Ford owners have fitted up their own electric lighting systems by procuring fittings available on the open market. The writer desired to use electric side and tail lamps instead of the oil lamps regularly furnished, but owing to the warning of the manufacturer of the car, no attempt was made to utilize the magneto current for this purpose. A 6-volt, 80-ampere hour storage battery was installed under the rear seat to furnish current. This proved to be thoroughly practical as it was only necessary to charge the battery 'once a month. A number of fittings, which are illustrated as they may be of value to the Ford owner who contemplates fitting electric lighting, are shown at Fig. 48. The application of a simple fitting to convert the square oil side light to an electric side light is shown at A. This has an advantage inasmuch as the oil burner may be used in event of failure of the source of current. The same fitting may be applied to the tail lamp. At B, a very compact side light is shown. The tail light is practically the same design except that it is smaller and has a red lens instead of the white glass in the door. If electric current from storage battery is provided small fixtures as shown at C may be used for illuminating the speedometer dial while portable search lights such as shown at E and D will assist in locating engine trouble and repairing tires after dark. The trouble light shown at D has a cigar lighter attached which would be found very convenient by the smoker. The combination of tail lamp supporting bracket and number plate holder shown at F is also very practical fitting.

A section through a typical high-grade electric head light is shown at G. This is supplied in cases where the motorist does not wish to use a make-shift reflector to convert the gas lamps. A 6-volt, 80-ampere battery will furnish enough current to operate two electric side lights, two moderate power head lights and a tail lamp for periods ranging from two weeks to a month without recharging, the service rendered being of course dependent upon the amount of night riding done. For those who wish only to use the side lamps and tail lamps of the electrical form, special low voltage bulbs may be obtained that will burn very satisfactorily on dry cell battery current. These bulbs are not sufficiently powerful for head lights, however, so some motorists fit electrical head lights taking the current from the magneto and depend on the dry battery only for the side and tail lamps. Twelve dry cells wired in series-multiple combination in which three sets of four joined in series are wired in multiple, will form a practical battery for use with the low voltage bulbs. A simple wiring diagram showing the method of coupling three lamps with the controlling switch and battery wired in is shown at Fig. 49.

The usual method of wiring head lights is to run a wire from the magneto terminal to a one point switch on the dash, from the switch to one side of the left head light double contact bulb. The head light bulb is then joined to its neighbor, the free terminal of which is grounded to the frame side member. This means that the lamps are wired in series, this being done to permit the use of six or seven-volt lamps which are a standard, easily procured size and at the same time insures against burning them out due to excess voltage generated at high engine speeds. If the head lights are connected to the storage battery they should be wired in multiple instead of series, just as the side lamps are in Fig. 49. Separate switches should be provided for the head lights, side light, and tail lamp circuits, inasmuch as it is not necessary to use all lamps at the same time. While the tail light must be kept burning at all times, it can be used in connection with the side lamps for city driving and these can be extinguished and only the head lights used for cross country work.

Tools and Supplies for Pneumatic Tire Restoration. — It has been previously stated that one of the chief disadvantages of pneumatic tires has been their liability of failure by puncturing the outer casing and penetrating the inner tube and thus providing a means for escape of the compressed air in the inner tube. The life of a pneumatic tire is decidedly uncertain and will

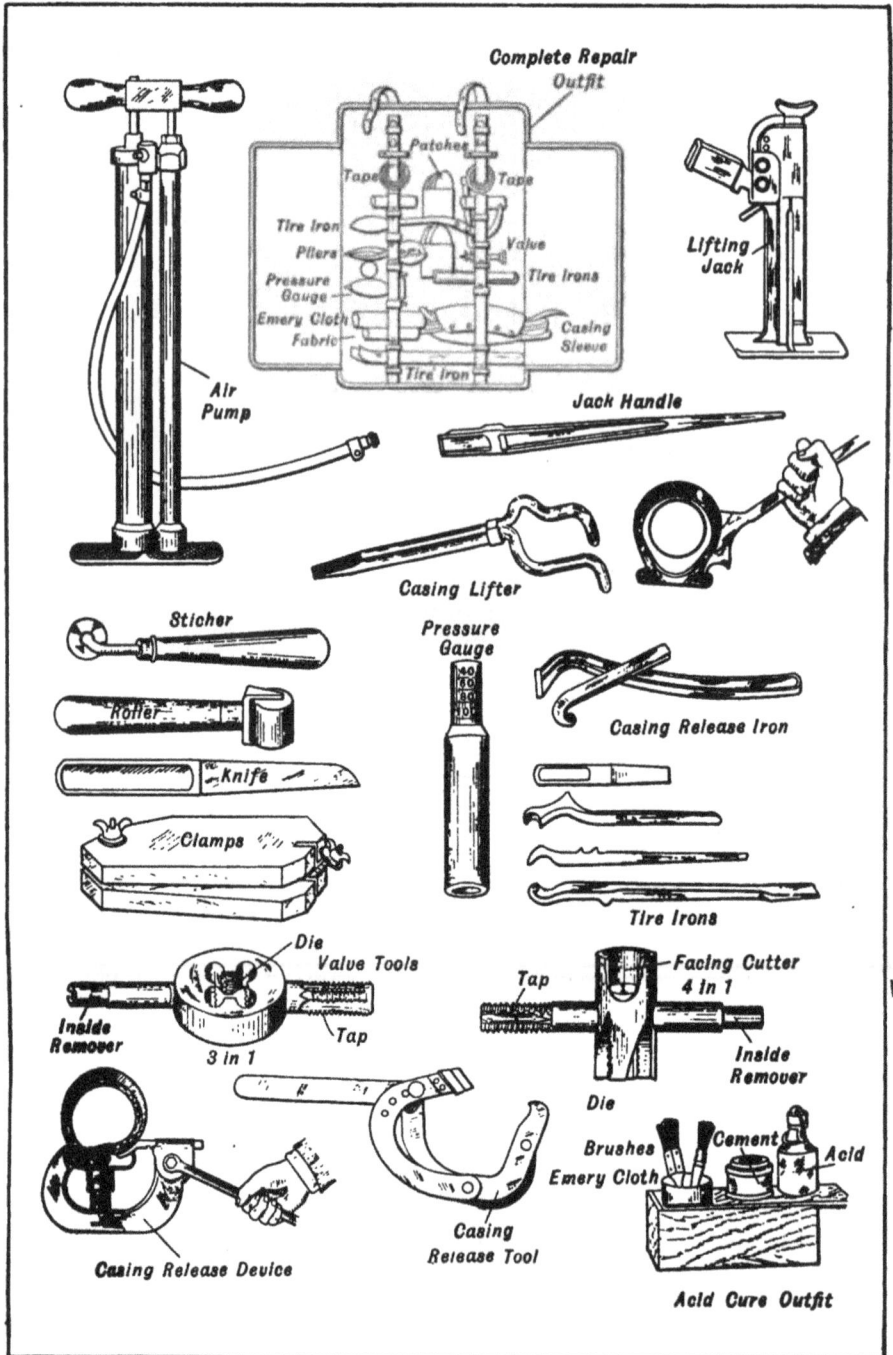

Fig. 50.—Tools and Supplies for Pneumatic Tire Maintenance, Application and Repair.

depend on many factors outside of those of purely natural wear. There have been cases where outer casings have given satisfactory service for seven or eight thousand miles, but these instances have been exceptions rather than the rule. It is the opinion of most motorists who have had practical experience that if an ordinary set of shoes will give a service averaging four thousand miles that they are equal to the demands made upon them and that they are satisfactory. It may be stated that tires will last longer on light cars like the Ford than on heavy ones and the service obtained from tires fitted to automobiles driven at low and moderate speeds will be much greater than that obtained from tires fitted to high speed vehicles. There is also a personal element which must be taken into consideration, and that is the way that the car is driven and the care taken of the shoes and inner tubes. It is necessary, therefore, in all cars using pneumatic tires to carry a certain amount of equipment for handling and repairing these on the road. A typical outfit is shown at Fig. 50, this supplementing two spare outer casings, and two extra inner tubes for replacement purposes. Included in the repair outfit are a blow-out sleeve, a number of patches and an acid-cure vulcanizing outfit for applying them. Tire irons are provided to remove the casing from the rim; the jack is used to raise the wheel of the vehicle on which the defective tire is installed from the ground and make it possible to remove the tire completely from the wheel. The air pump is needed to inflate the repaired tube or the new member inserted to take its place. Talcum powder is provided to sprinkle between the casing and the tube to prevent chafing or heating, while the spare valves and valve tool will be found useful in event of damage to that important component of the inner tube. As it is desirable to inflate the tires to a certain definite pressure, a small gauge which will show the amount of compression in the tire is useful.

The outfit shown may be supplemented by other forms of vulcanizing sets and by special tire irons to make for easier removal of the outer casing. Tire irons vary in design, and most makers of tires provide levers for manipulating the casings, which differ to some extent. A set of tire irons, such as would be needed with a clincher tire equipment, could be selected from the forms shown at Fig. 50. That shown near the gauge is utilized to loosen the clincher bead from under the rim should it become rusted in place. After the shoe has been loosened from the rim flange, levers of the form shown at or below it would be inserted under the bead in order to lift it over the rim. Two or more of these levers are necessary, the long ones being more easily operated than the short ones. The length of the lever provided will depend entirely upon the size of the tire to be removed. Motorists, as a rule, should carry one of the releasing levers shown, two of the short members depicted and one longer lever. The latter may be a combination form, which can be used as a jack handle as well as a tire iron, and then it is not necessary to carry a jack handle in the equipment. The flattened ends are generally employed for prying the bead from the clincher rim, and when this has been done and sufficient space exists between the bead and the rim to insert the curved end of the large levers, considerable leverage is obtained and the bead may be lifted over the clincher rim without undue exertion. The object of rounding the corners, and of making the working portions as broad as possible, is to reduce the liability of pinching the inner tube, which would be present if the irons had sharp edges.

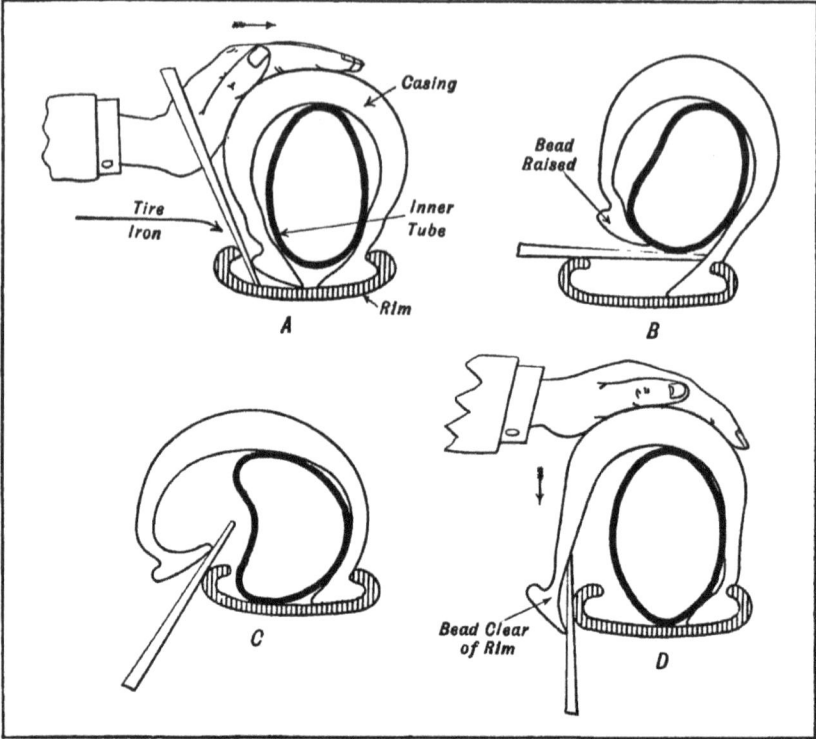

Fig. 51.—Showing Method of Releasing Clincher Casing from Rim. A.—Inserting the Tire Iron. B.—Raising the Bead. C.—Working the Clincher Bead Over the Edge of the Rim. D.—Method of Guiding Bead Over the Rim.

The tire repair material is sometimes carried in a special case, as shown at top of Fig. 50, this consisting of all parts necessary to make temporary repairs, to be considered in proper sequence. This outfit is sometimes supplemented by other special tools. A knife is needed to cut the rubber, trim patches, etc. The stitcher and roller are useful in rolling the patch after it has been cemented to the tire to insure adhesion of the patch firmly against the damaged portion of the tube while the cement is drying. Some motorists carry a small flame-heated vulcanizer in order to effect more permanent repairs than would be possible with the simple patching processes in which only the adhesive powers of dry cement are available.

Tire Manipulation Hints. — In removing or replacing outer casings considerable care must be exercised not to injure the shoe or pinch the inner tube. The first step is to jack up the wheel from which the defective tire is to be removed, thus relieving the wheel of the car weight.

The valve inside is then unscrewed in order to allow any air that may remain in the tube to escape, and then the lock nuts on the valve stem are removed so that this member may be lifted to release the clincher beads from the rim channels. If the tire is stiff or has not been removed for some time, a special iron is utilized to loosen the edges and the beads are pushed clear of the clincher rim.

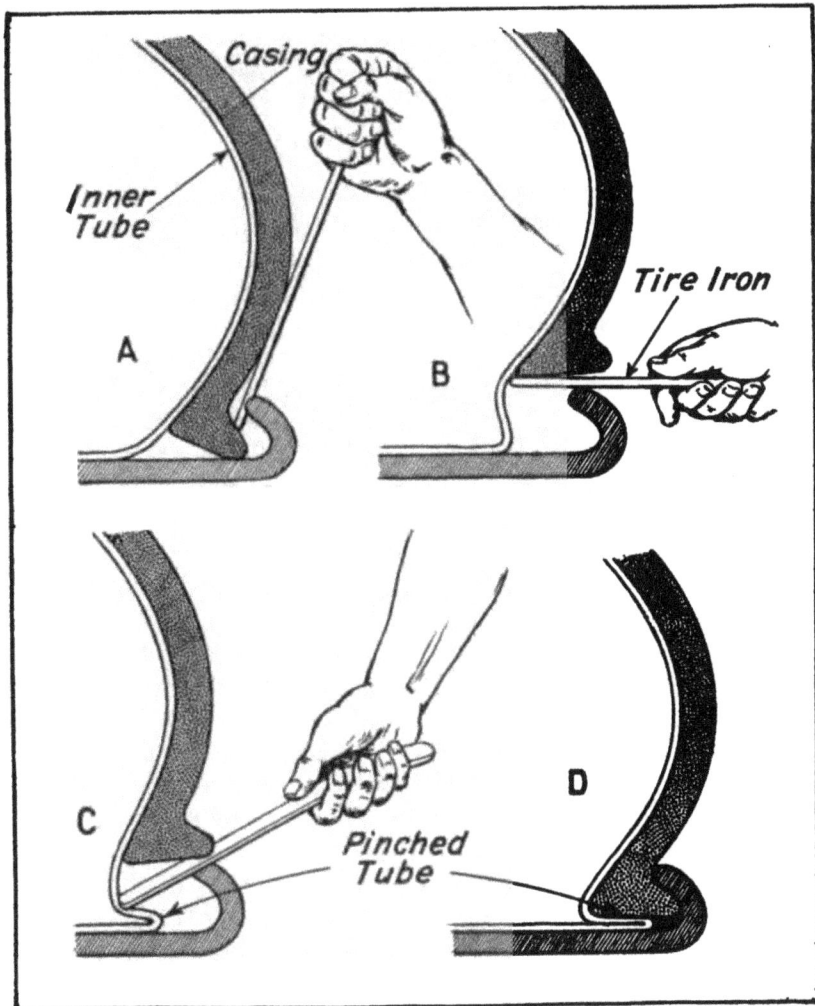

Fig. 52.—Proper Methods of Handling Tire Irons in Removing and Replacing Outer Casings on Clincher Rims at A and B. How Inner Tube May Be Pinched if Tire Iron is Carelessly Manipulated at C. Inner Tube May Be Pinched if Placed in Casing Without Being Partially Inflated as at D.

When the casing has been loosened on one side, a flat tool, as shown at Fig. 51, A, is inserted under the loose bead to act as a pry or lever to work the edge of the casing gradually over the rim. Very long levers are necessary to handle new stiff tires, and unused casings are particularly hard to move. The shorter irons may be employed on shoes which have been used for some time and which are more pliable than the new ones. Two of the levers are generally used together, one being kept under the loosened edge of the bead, while the other is used to force the bead over the edge of the rim. When the outside edge of the bead has been forced over the rim at all points the inner tube is lifted from the rim and is pulled out of the shoe. The start at removing is made at the point diametrically opposite the valve stem. When this portion has been pulled clear of the rim and out of the casing it is not difficult to pull the rest of the tube out and finally lift the valve stem out of the hole through which it passes in the wheel felloe, and take the inner tube entirely off the wheel. If the casing demands attention, or if a new shoe is to be used, the inside bead is worked over the channel of the clincher rim in just the same manner as was done with the outside bead, and after a start has been made and a portion of the inside bead forced over the rim there will be no difficulty in slipping the entire shoe from the wheel.

Applying a tire is just the reverse to removing one. The first operation is to place the inner bead of the tire in place in the center of the rim by forcing it over the outside flange. This is done gradually, and in order to force the remaining portion of the shoe it may be necessary to use long levers when the greater part of the casing has been applied. The next step is to work the shoe gradually toward the inner channel of the rim, then to insert the air tube. The inner tube is replaced after it has been partially inflated by putting the valve stem in first and then inserting the rest of the tube, being careful not to pinch it under the beads.

After the inner tube has been put in place, the outer bead of the tire is worked over the edge of the rim channel. Care must be exercised to insure that the inner tube will not be pinched by the sharp edges of the tire levers, as at Fig 52, C. The object of partially inflating the inner tube is to distend it so there are no loose or flabby portions that are liable to catch under the tire bead when this is being forced in place over the wheel rim, as at Fig. 52, D. The conventional method of inflating tires by using a foot pump does not always insure that they will receive adequate inflation, and when a pump is employed it is imperative that some form of gauge be provided that will register the amount of pressure inside of the tire in order that it will reach the figure recommended by the tire makers. Different methods of tire inflation have been devised which eliminate the necessity of using manually operated pumps. Obviously, a simple expedient would be to provide a small power-driven pump that could be actuated by any convenient mechanical connection with the engine or a spark plug pump. Another method is to use an air bottle, which is a steel container in which air is stored under great pressure. The air is compressed to such a point that a tank less than two feet long and six inches in diameter will furnish sufficient air to inflate seven or eight rear tires or twelve front ones. The tanks may be exchanged at small cost, when exhausted, for new containers holding a fresh supply of air. In some tanks, gases of various kinds under high pressure are used, and the motorist may obtain these on the same basis as air bottles are supplied.

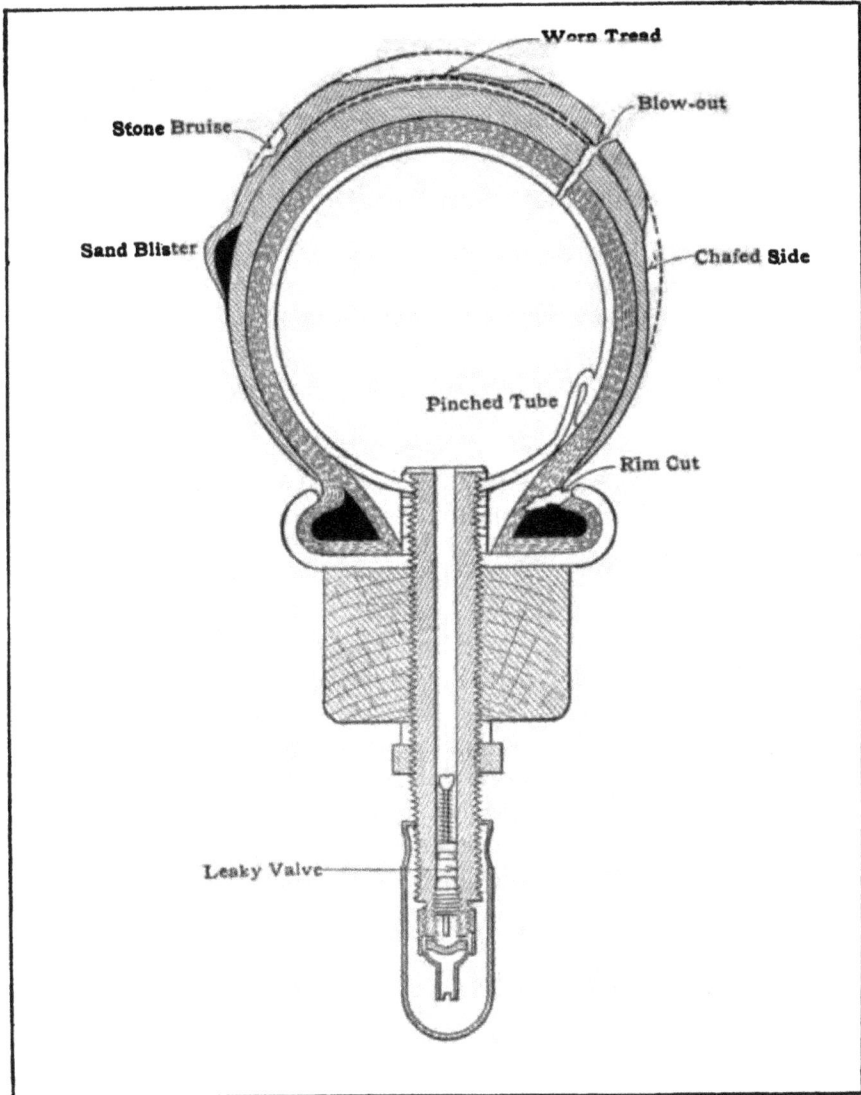

Fig. 53.—Cross Section of Typical Clincher Tire Showing Defective Points that Demand Attention When Restoring Tires to Proper Condition.

All devices of this character are fitted with gauges to indicate the amount of pressure in the tire, and to prevent overinflation. If a tire is not properly inflated the shoe will be liable to various kinds of road damage and will be easily punctured, while if the pressure is too high the shoe is liable to "blow-out" at any weak point in the structure. A tire-pressure gauge is a very necessary article of equipment in any car, and its proper use when blowing up tires will insure the best possible results if the schedule recommended by the tire manufacturers is adhered to. Ford front tires should be inflated to 60 pounds, the rear ones to 70 pounds pressure. The rule is 20 pounds for every inch of tire width.

Tire Repair and Maintenance. — The common causes of tire failure that the motorist is apt to encounter are shown at Fig. 53. The most common is natural wear of the tread portion of the tire. The rubber compound in contact with the road surface wears away in time, and the fabric layers ' which constitute the breaker strips are exposed! The shoe is weakened and any sharp object in the road is apt to penetrate the weakened case and puncture the inner tube. If a number of the layers of fabric comprising the body of the shoe are cut this constitutes a weak place in the casing and a blow-out will result, because the few layers of fabric remaining do not have sufficient strength to resist the air pressure. A stone bruise is caused by the removal of a portion of the rubber tread by a sharp stone, piece of glass, etc., and is much more serious than a puncture, because it removes some of the tire, whereas in ordinary cases of puncture a sharp object merely penetrates the casing. A sand blister is produced by sand or grit from the road working into a space in the tire between the tread and the fabric body through some neglected incision or bruise. The side of the tread is often chafed by running the tires against curbstones or by driving in a car track. Rim cutting is generally caused by insufficient inflation, which permits the rim to cut into the tire and thus tends to sever the bead from the side of the shoe.

The chief inner tube trouble is penetration of the wall by some sharp object, or the folding over of part of the tube walls when the tire was applied. The parts of the check valve sometimes give trouble and the valve leaks. In cases of valve trouble it is usually cheaper to replace the valve inside than it is to attempt to fix it. Some of the causes of valve leakage are hardening of the rubber washer, bent stem, which prevents the valve from seating properly, or a particle of dust or other foreign matter which would act to keep the valve from closing the air passage positively.

The most serious condition that a motorist will meet with is a "blow-out," and usually only temporary repairs can be made on the road. The common methods of restoring a defective outer casing are depicted at Fig. 54. In this, an inner sleeve, which is composed of a number of plies of fabric, is placed between the inner tube and the broken portion of the outer casing to prevent pinching of the inner tube by the jagged edges of the cut. To strengthen the casing from the outside, an outer shoe or gaiter made of leather is laced around the shoe. The object of using both inside and outside reenforcing members in combination is to not only strengthen the weak outer casing but by providing an outer shoe dirt is kept from working into the tire.

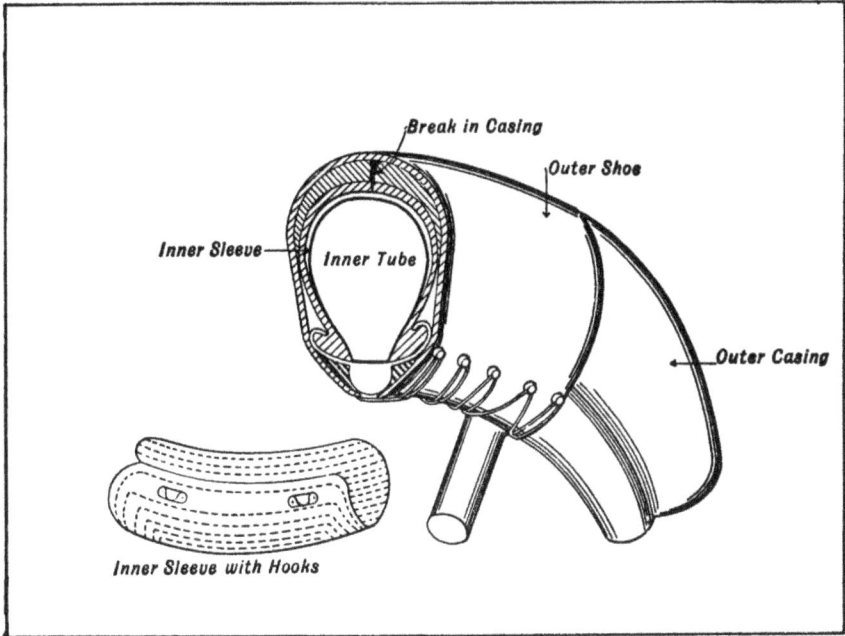

Fig. 54.—Showing the Application of Inner and Outer Casing Sleeves as a Temporary Repair for Ruptured Outer Casings.

Punctured inner tubes may be temporarily repaired by using a cemented surface patch. The first step necessary is to clean the surface of the tube very thoroughly with gasoline and then to rough up the surface of both patch and that portion of the tube surrounding the hole with a wire scratch brush or with sandpaper. After the surfaces are properly cleaned and roughened the patch and the tube are coated with suitable patching cement, which is allowed to become thoroughly dry before the second coat is applied. The second coat is allowed to become "Hacky," which expresses a condition where the cement is almost dry and yet still possesses a certain degree of adhesiveness. This requires about ten minutes' time on an average. The patch is applied to the cemented portion of the tube and the whole is clamped firmly together to secure positive adhesion while the cementing medium is drying. Patches should always be of sufficient size to cover the damaged portion and at the same time have about threequarters of an inch or more of the patch at all sides of the orifice.

Very satisfactory repairs to both inner and outer casings of a permanent nature can be made by using small portable vulcanizers, which may be heated by either electricity, steam or by burning gasoline in them. When these are used a special vulcanizing cement is necessary, and uncured rubber stock must be used for patching or filling openings caused by punctures or blowouts. The patch of raw material is applied to the cemented surface of tube or casing and the vulcanizer heated to the proper temperature. The heat of the vulcanizer causes the rubber of the patch to unite perfectly with the old material and forms an intimate bond. In vulcanizing the most important precaution is to maintain a proper temperature. Too great a degree of heat will burn the rubber, while a proper cure cannot be effected if the temperature is too low. The temperatures recommended for vulcanizing vary from 250 to 325 degrees F. The lower degree of heat is used for working material that has been previously cured, while the higher temperature is recommended for new rubber.

The rules to secure satisfactory operation from pneumatic tires may be easily summed up. In the first place, it is imperative that the tires be inflated to the pressures recommended by the manufacturers. The tires should be kept clean and free from oil or grease, because the oleaginous substances used for lubrication very quickly attack rubber compounds and cause crumbling and rapid deterioration. Oil or grease should be wiped off as soon as noticed and the tire cleaned by application of gasoline. Any small cuts or openings in the tire that may permit water to enter or sand to work between the fabric and the tread will cause trouble in time and should be filled with rubber compound as soon as noticed. One should be careful in driving not to apply the brakes too suddenly, because this will lock the wheels and wear the tires flat in spots very quickly. Care should be taken not to drive in car tracks, and when highways do not have the proper surface they should be negotiated very carefully to avoid cutting the casings.

Typical Special Tool Equipment. — The makers of all the popular cars, especially those that are produced in any quantity, furnish special tools for the use of their repair men or for those employed in the service stations of the agents. As an example of the special tools that it is possible to obtain the assortment used by repairmen of Ford Model T cars is shown at Fig. 55. The device at A is a wheel puller designed to go on the hub in place of the hub cap. The tools shown at B, C, D, F, G and D2 form part of the regular tool equipment. The box wrench at E is also furnished with each car, and is a hub cap wrench having one end formed to fit the slabbed portion of the front wheel bearing adjusting cone lock nut.

A valve spring lifter is shown at H, while a valve seat reamer is shown at I. The valves are turned while grinding by the special brace S, which can be used on all of the valves except the one on the rear cylinder, which is immediately under the dash board. To turn this valve the special wrench shown at L is provided. A special T handle socket wrench for handling ⅝" nuts or bolt heads such as used on the rear construction and various other points is shown at J. AT handle screw driver for the set screws which are employed in retaining the camshaft bearings is shown at O. The puller shown at E is for removing the cam gear from the camshaft, while that at Q is a transmission clutch puller. The brace shown at E is a special socket wrench for 3/8" bolt nuts. The brace shown at T is employed for removing the magnet retaining screws in the magneto assembly. The tire irons at A-2, the tool roll at B, the pump at D-1 and the spark plug socket wrench at D-2 are all parts of the regular tool equipment furnished with each car.

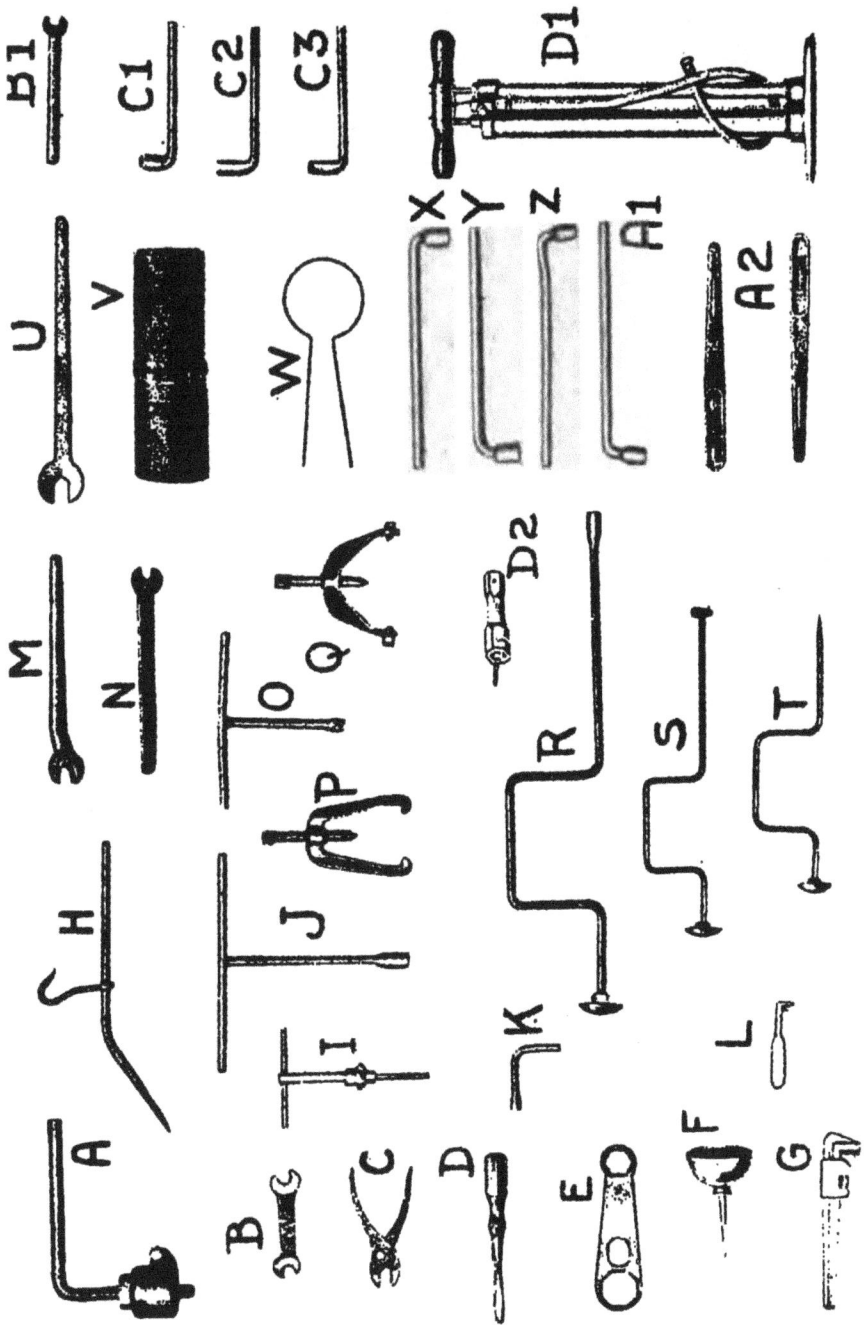

Fig. 55.—Special Tool Outfit for Repairing Ford Cars.

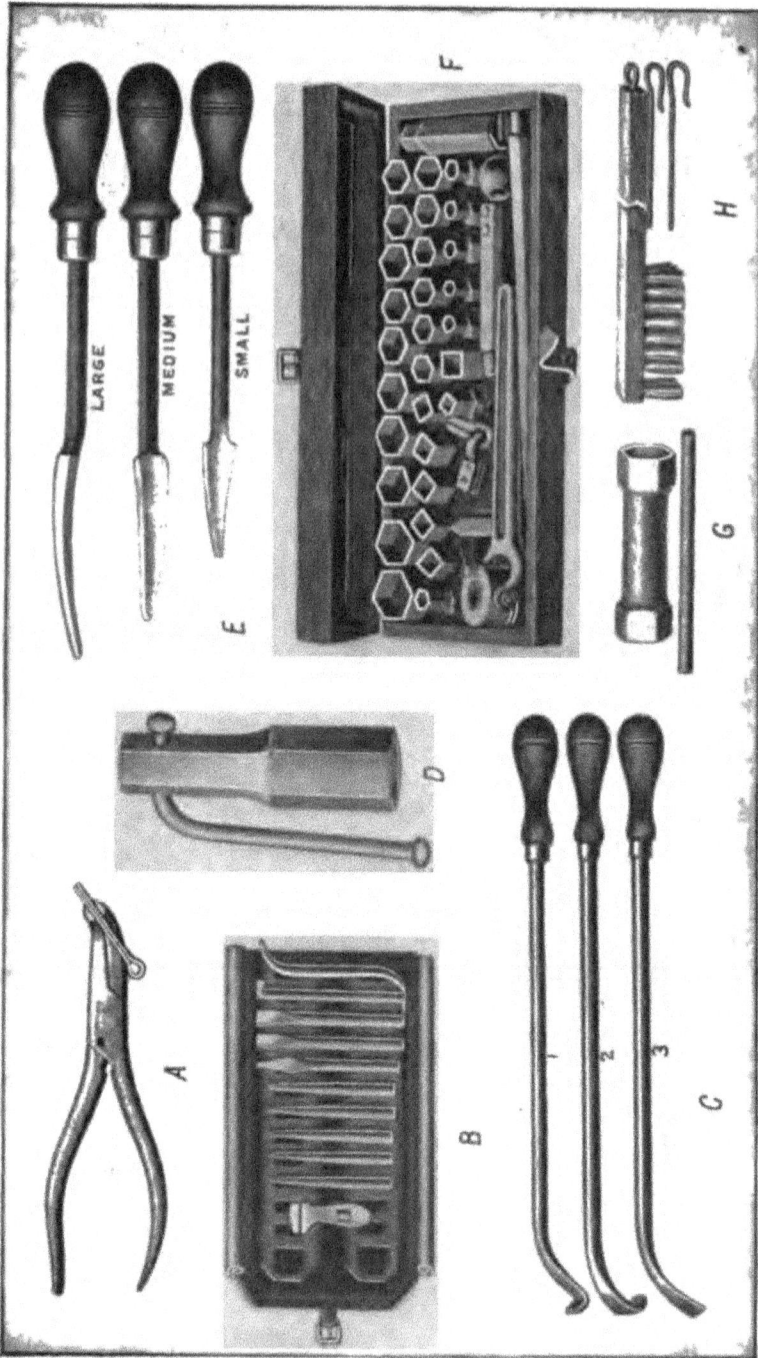

Fig. 56.—Tools That Will Be Valuable When Overhauling or Repairing Ford Automobiles. A.—Special Pliers for Removing and Inserting Split Pins. B.—Chisel and Punch Set. C.—Carbon Scrapers. D.—Socket Wrench. E.—Bearing Scrapers. F.—Batchet and Socket Wrench Set. G.—Double End Box Wrench. H.—Spark Plug Brush.

The simple fitting shown at W is a piston ring compressor employed to compress the rings in the piston grooves to facilitate easy assembly in the

104

cylinder block. A number of special socket wrenches are shown at X, Y, Z, A-1, C-1, C-2 and C-3. These are all intended for use on the various fastenings used in holding the parts together. For example, that at X is a socket wrench for the crank shaft main bearing bolt nuts. That at Y is for 3/8" bolt heads or nuts. The wrench at Z is for removing the cylinder head retaining bolts. The wrench for removing the pinion drive shaft housing retaining stud nuts is shown at C-1, this being used for 3/8" nuts. The rear axle housing bolt nut wrench is shown at C-2, while the form outlined at C-3 is similar to that shown at C-1, except that it fits 11/32" nuts. The special end wrench at M is for the flywheel retention cap screws, that at U is for removing the large cam gear lock nut, while that at B-1 is a regular open end wrench for 3/8" nuts. The prices on these tools are so low that it is cheaper to purchase from the factory than to attempt to make them.

A Typical Engine Stoppage Analyzed. — Before describing the points that may fail in the various auxiliary systems, it will be well to assume a typical case of engine failure and show the process of locating the trouble in a systematic manner by indicating the various steps which are in logical order and which could reasonably be followed. In any case of engine failure the ignition system, motor compression and carburetion should be tested first. If the ignition system is functioning properly one should determine the amount of compression in all cylinders, and if this is satisfactory the carburetting group should be tested. If the ignition system is working properly and there is a decided resistance in the cylinders when the starting handle is turned, proving that there is good compression, one may suspect the carburettor.

It is possible that the inlet manifold may be loosely held or cracked, that an exhaust valve is stuck on its seat because of a broken or bent stem. Failure of the camshaft drive because the teeth are stripped from the engine shaft or cam-shaft gears; or because the key or other fastening on either gear has failed, allowing that member to turn independently of the shaft to which it normally is attached, is very rare. The gasoline feed pipe may be clogged or broken, the fuel supply may be depleted, or the shut-off cock in the gasoline line may have jarred closed. The gasoline filter screen or sediment bulb may be filled with dirt or water, which prevents passage of the fuel. The defects outlined above, except the failure of the gasoline supply, are uncommon, and if the container is found to contain fuel and the pipe line to be clear to the carburettor, it is safe to assume that the vaporizing device is at fault. If fuel continually runs out of the mixing chamber the carburettor is said to be flooded. This condition results from failure of the shut-off needle to seat properly or from a gasoline soaked cork float. It is possible that not enough gasoline is present in the float chamber. If the passage controlled by the float needle valve is clogged or if the float was badly out of adjustment, this contingency would be probable. When the carburettor is examined, if the gasoline level appears to be at the proper height, one may suspect that a particle of lint, dust, fine scale or rust from the gasoline tank has clogged the bore of the fuel passage.

On cranking the motor slowly over by hand, if one cylinder has poor compression while all others have good compression the trouble may be due to a defective condition either inside or outside of that cylinder. The external parts may be inspected more easily, so the following should be looked for: a broken valve, a warped valve head, broken valve springs, sticking or bent valve stems, dirt under valve seat, leak at cylinder head packing or spark plug gasket, cracked cylinder head (rarely occurs), leak through cracked spark plug insulation, valve plunger stuck in the guide, lack of clearance between valve-stem end and top of plunger. The faulty compression may be due to defects inside the motor. The piston head may be cracked (rarely occurs), piston rings may have lost their elasticity or have become gummed in the grooves of the piston or the piston and cylinder walls may be badly scored by a loose wrist pin or by defective lubrication. The Ford motor is a type with a separately cast head and it is possible that the gasket or packing between the cylinder and combustion chamber may leak, either admitting water to the cylinder or allowing compression to escape. This may be corrected by tightening all the retaining bolts firmly to seat the head against the gasket.

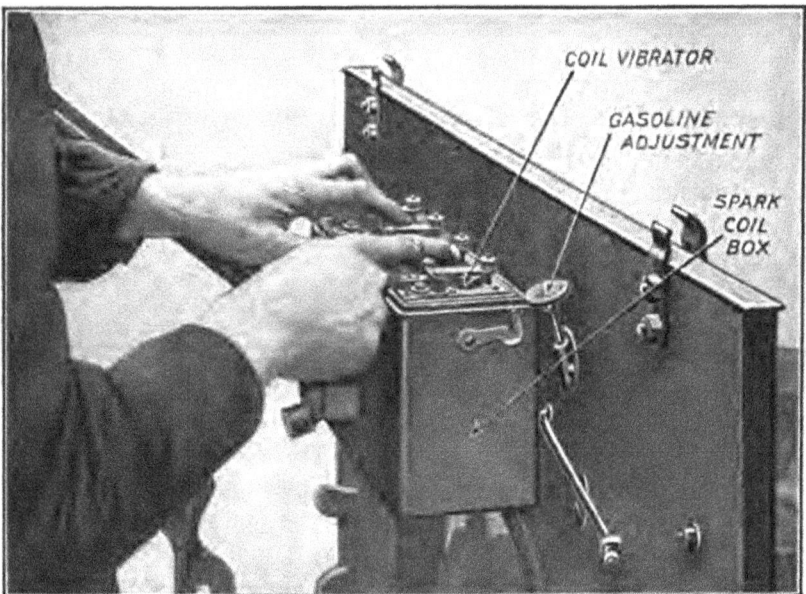

Fig. 57.—Method of Testing Regularity of Engine Action by Holding Down Coil Vibrators.

Conditions That Cause Failure of Ignition System. — If the first test of the motor had showed that the compression was as it should be and that there was no serious mechanical defects and there was plenty of gasoline at the carburettor, this would have demonstrated that the ignition system was not functioning properly. If a battery is employed as an auxiliary supply of

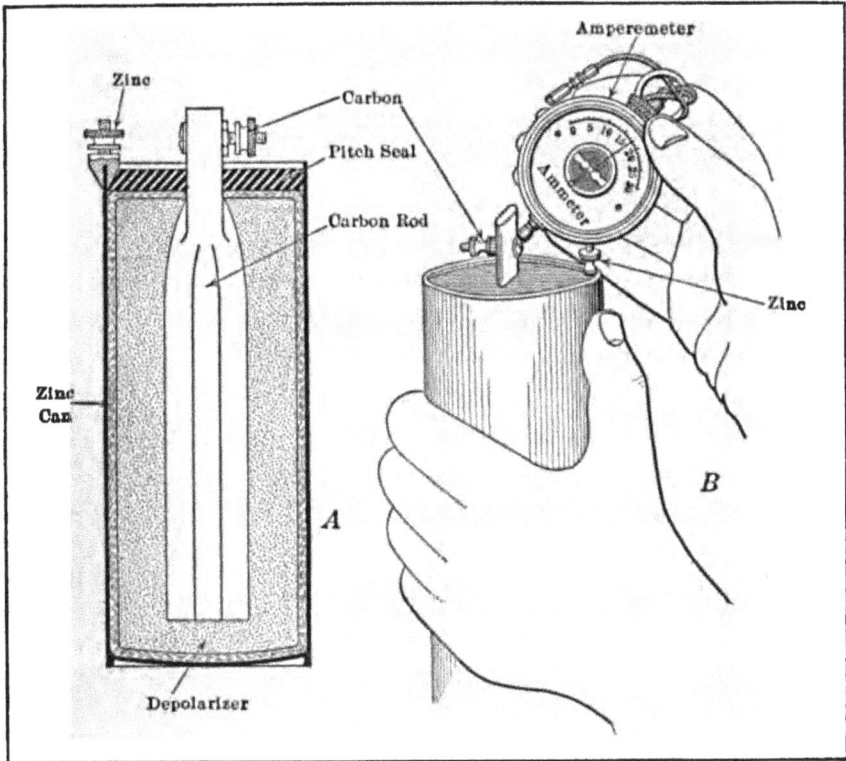

Fig. 58.—Sectional View of Dry Cell Showing Interior Construction at A and Method of Testing Current Output With Amperemeter at B.

current, the first step is to take the spark plugs out of the cylinders and test the system by turning over the engine by hand, with the switch circuit closed. If there is no spark in any of the plugs, this may be considered a positive indication that there is a broken main current lead from the battery, a defective ground connection, a loose battery terminal, or a broken cell connector. If none of these conditions are present, it is safe to say that the battery is no longer capable of delivering current. If there is no spark at the plugs, but the spark-coil vibrator functions properly, this shows that the primary wiring is as it should be and that the fault must be looked for in either the wires comprising the secondary circuit, or at the plugs. The spark plugs may be short circuited by cracked insulation or carbon and oil deposits around the electrodes. The secondary wires may be broken or have defective insulation, which permits the current to ground to some metal part of the frame or motor. The battery strength should be tested with a volt or ampere meter to determine if the voltage and amperage is sufficient. Storage battery capacity is usually gauged by measuring the voltage, while dry cells are judged by their amperage. A storage battery should show over two volts per cell or 6.5 volts in a three-cell group, while dry batteries that indicate less than six amperes per cell are not considered reliable or satisfactory for ignition service. Look also at the magneto contact plunger to make sure it is clean and not gummed up or held away from its seat by a small shred from the brake lining. If there is no vibration at the coil tremblers the trouble may be due to weak current source, broken or loose magneto wire, broken timer wires, or defective connections at the vibrator or commutator contact points. The electrodes of the spark plug may be too far apart to permit a spark to overcome the resistance of the compressed gas, even if a spark jumps the air space when the plug is laid on the cylinder. In some cases the trouble has been due to the switch, which did not contact properly. In others, the spring keeping the timer case in place loosened and the timer case was not making proper contact with engine. It should always be in firm contact with the engine base to secure positive electrical contact between timer roll carrier and segments in timer case.

If the motor runs intermittently, i.e., starts and runs only a few revolutions, aside from the conditions previously outlined, defective operation may be due to seizing between parts because of insufficient oil or deficient cooling, too much oil in the crank case, which fouls the cylinders after the crank has revolved a few turns, and derangements in the ignition or carburetion systems that may be easily remedied. There are a number of defective conditions which may exist in the ignition group, that will result in "skipping" or irregular operation, and the following is the logical order in which the various points should be inspected: the parts which demand inspection oftenest are considered first; weak source of current due to worn-out dry cells or discharged storage batteries; weak magnets in magneto; dirt or gummed oil in timer casing, broken spring on timer roll carrier, or poor contact at magneto current collecting plunger. A dirty or cracked insulator at spark plug will cause short circuit, and can only be detected by careful examination. The following points should also be checked over when the plug is inspected: Excessive space between electrodes, points too close together, loose central electrodes, or loose point on plug body, soot or oil particles between electrodes, or on the surface of the insulator.

How to Locate "Skipping" Cylinder. — If irregular engine action is due to poor ignition and the trouble seems localized to some particular cylinder, it is very easy to locate the member at fault by holding down the coil vibrators as shown at Fig. 57. The engine is run with a late spark and with the throttle about half open and three of the vibrators are held down at a time, leaving the engine running on one cylinder only. It may often be found that three of the cylinders are functioning properly and that only one is at fault, which in this case means that the trouble is present either at the spark plug or to some condition in that cylinder other than faulty ignition. The engine will run on one cylinder when three of the coil vibrators are held down, the object being to discover the weak or faulty cylinder by putting those that are running properly out of action temporarily. When all three vibrators that belong to the cylinders that are working properly are held down the engine will not run with the vibrator serving the cylinder that is at fault free. Some repair-men hold down only two of the vibrators and let the engine ran on the re-maining one or two cylinders. This is in cases where the mixture adjustment may be faulty enough so that one cylinder will not suffice to keep the engine in operation.

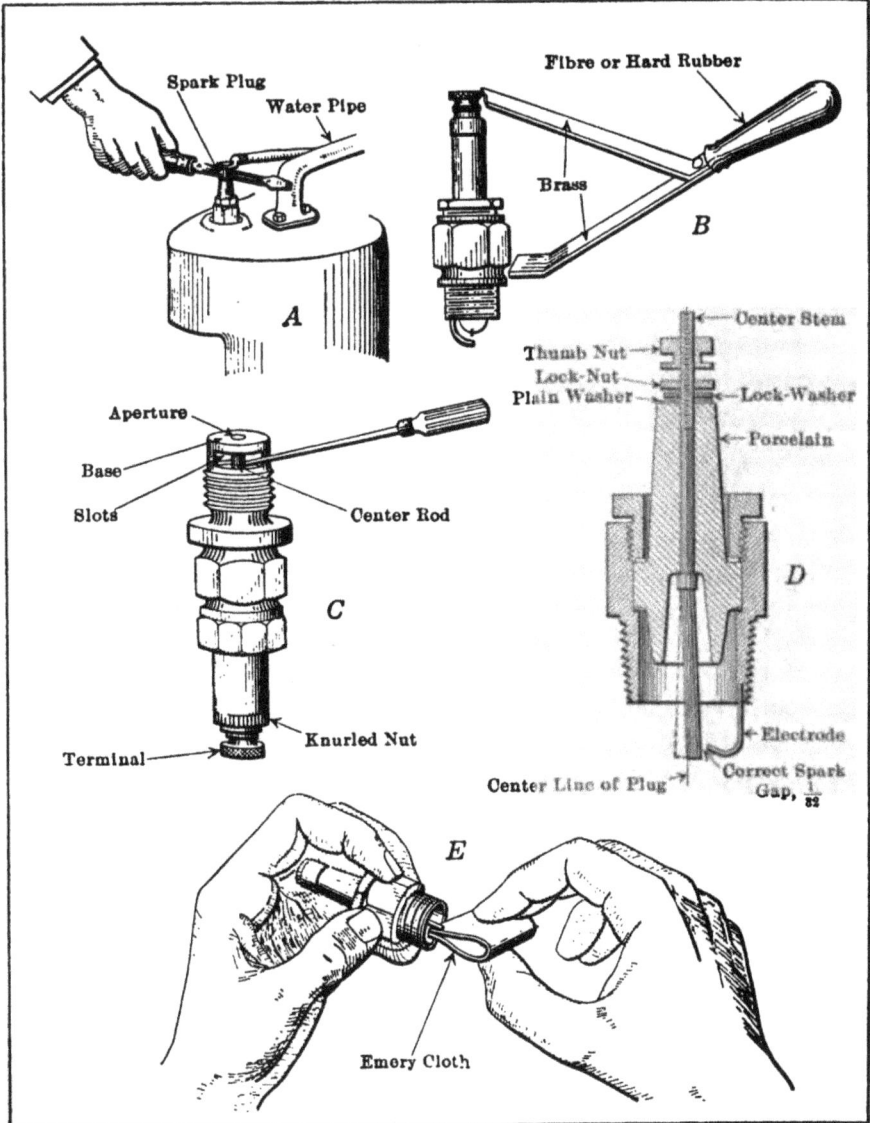

Fig. 59.—Showing Methods of Adjusting Air Gaps in Spark Plugs.

Spark Plug Faults. — After the "missing" cylinder has been located either by the method of holding down the vibrators or by short circuiting each of the cylinders in turn by using a screw driver as indicated at Fig. 59, A to connect the insulated terminal of the spark plug to some metal portion of the engine, the next step is to remove the spark plug and examine it carefully for various faults. On some cars using high tension magneto ignition and on some Fords that have been fitted over in this way by their owners it is not easy to detect irregular engine action, as no vibrators are used that may be utilized as indicators of correct engine action. In this case the simple tool shown at Fig. 59, B may be used. This consists of two strips of brass riveted together and fastened into a fiber or hard rubber handle as shown. These make a very effective short-circuiting medium and entirely eliminate the danger of shock present when a screw driver is used for this purpose. If the plug is dirty the point should be cleaned with an old toothbrush dipped in gasoline or with a special spark plug brush made for the purpose as outlined at Fig. 56, H.

If the plug is extremely dirty it should be taken apart, which should be easily done by removing the porcelain retaining bushing that fits the main steel shell and then taking the porcelain out of the shell member. Care should be taken not to drop any of the copper asbestos gaskets used on both sides of the porcelain shoulder. The carbon deposits, which are usually of an oily nature, may be easily removed from the porcelain or shell interior with the small blade of a jack knife, but care should be taken not to scratch the glazed surface of the porcelain, as this permits the insulator to absorb oil. All parts of the plug should be thoroughly washed with gasoline and wiped dry with a clean rag. When reassembling the plugs care should be taken that the bushing holding the porcelain in place is not tightened too much, as this may crack the insulator. It is also important that the distance between the spark points should be no more than 1/32", which is about the thickness of a worn dime. Some plugs have a center terminal, as shown at Fig. 59, C. These are easily handled by pushing the center rod over until it bears the proper relation to the aperture in the plug base.

The standard form of plug used on Ford cars is shown at Fig. 59, D. With this construction the air gap is maintained at the proper point by moving the center stem as conditions demand. A cracked insulator must be replaced with a new one, and if the bent electrode is found loose in the plug shell this should be firmly retained in place by jamming a portion of the metal surrounding it toward the wire with a sharp prick punch. The method of using a doubled over piece of fine emery cloth to brighten the plug points to insure the easy passage of the ignition spark is clearly shown at Fig. 59, E. The spark plug should be tested after it is assembled by connecting it to the spark plug wire and laying it on the cylinder in such a way that only the steel shell is in contact with the metal parts of the engine. The engine is then cranked over by hand with the switch on the coil box in the battery position if these members are provided and the plug watched to see if a spark occurs between the points at regular intervals. If the spark plugs foul up quickly the trouble is usually due to too much lubricating oil in the engine, worn piston rings, which allow the oil to reach the combustion chambers, the use of poor quality oil or too rich fuel mixture.

Faults in Other Ignition System Parts. — When testing a dry battery, the terminals should be gone over carefully to make sure that all terminal nuts are tight and that there are no loose or broken connectors. The wiring at the coil, timer, and switch should be inspected to see that all connections are tight and that the insulation is not chafed or cracked. Defective insulation will allow leakage of current, while loose connections make for irregular operation. In testing a storage battery care should be taken to remove all the verdigris, or sulphate, from the terminals before attaching the testing wires. The strength of the Ford magneto can be determined by the Hoyt magnetometer, a device specially made for this purpose. In the Ford system a vibrator coil is employed, so the trembler platinum contact points should be examined for pits, or carbonized particles that would interfere with good contact. If defective, they should be thoroughly cleaned and the surfaces of the platinum point on both vibrator spring and adjusting screw should be filed smooth to insure positive contact. The tension of the vibrator spring should not be too light or too heavy, and the vibrator should work rapidly enough to make a sharp, buzzing sound when contact is established at the timer. The adjusting screw should be tight in the vibrator bridge, and when proper spring tension is obtained the regulating screw should be locked firmly to prevent movement, if an automatic lock is not provided, as on Heinze coils.

If the vibrator operates satisfactorily, but there is a brilliant spark between the vibrator points and a poor spark at the spark plug, one may assume that the coil condenser is punctured. Short circuits in the condenser or internal wiring of induction coils or magnetos, which are fortunately not common, can seldom be remedied except at the factory where these devices were made. If an engine stops suddenly and the defect is in the ignition system, the trouble is usually never more serious than a broken or loose wire. This may be easily located by inspecting the wiring at the terminals. Irregular operation or misfiring is harder to locate, because the trouble can only be found after the many possible defective conditions have been checked over, one by one.

Common Defects in Fuel Systems. — Defective carburetion often causes misfiring or irregular operation. The derangements of the components of fuel system that are common enough to warrant suspicion and the best methods for their location follows: First, disconnect the feed pipe from the carburettor and see if the gasoline flows freely from the tank when the shut-off valve is opened again. If the stream coming out of the pipe is not the full size of the orifice it is an indication that the pipe is clogged with dirt or that there is an accumulation of rust, scale, or lint in the strainer screen of the filter or that the sediment bulb is filled with foreign matter. It is also possible that the fuel shut-off valve may be wholly or partly closed. When the gasoline flows by gravity the liquid may be air bound in the tank on account of a plugged vent hole in filler cap.

If the gasoline flows from the pipe in a steady stream the carburettor demands examination. There may be dirt or water in the float chamber, which will constrict the passage between the float chamber and the spray opening, or a particle of foreign matter may have entered and clogged the fuel inlet. The float lever may bind on its guide pin, the needle valve regulating the gasoline inlet opening in bowl may stick to its seat. Any of the conditions mentioned would cut down the gasoline supply and the engine would not receive sufficient quantities of gas. The gasoline adjusting needle may be loose and jar out of adjustment. Air may leak in through the manifold, due to a porous casting, or leaky joints due to poor gaskets or loose retaining stirrups and dilute the mixture. Water or sediment in the gasoline will cause misfiring because the fuel feed varies when the water or dirt constricts the pipe bore.

It is possible that the carburettor may be out of adjustment. If clouds of black smoke are emitted at the exhaust pipe it is a positive indication that too much gasoline is being supplied the mixture, and the supply should be cut down by screwing down the needle valve and by making sure that the fuel level is at the proper height in the float bowl. If the mixture contains too much air there will be a pronounced popping back in the carburettor. When a carburettor is properly adjusted and the mixture delivered to the cylinders burns properly, the exhaust gas will be clean and free from the objectionable odor present when gasoline is burned in excess. If the muffler » cut-out has been provided by the owner the character of combustion may be judged by the color of the flame which issues from it when the engine is running with an open throttle after nightfall. If the flame is red, it indicates too much gasoline. If yellowish, it shows an excess of air, while a properly proportioned mixture will be evidenced by a blue flame, such as given by a gas stove burner.

Defects in Oiling and Cooling Systems. — While troubles existing in the ignition or carburetion groups are usually denoted by imperfect operation of the motor, such as lost power and misfiring, derangements of the lubrication or cooling systems are usually evident by overheating, diminution in engine capacity, or noisy operation. Over-heating may be caused by poor carburetion as much as by deficient cooling or insufficient oiling. When the oiling group is not functioning as it should the friction between the motor parts produces heat. If the cooling system is in proper condition, as will be evidenced by the height of the water in the radiator, and the carburetion group appears to be in good condition, the overheating is probably caused by some defect in the oiling system.

The Ford oiling system is so simple that practically the only condition that will cause defective oiling is lack of oil in the engine case or the use of poor grade lubricant. The use of dirty oil will also produce overheating and may result in trouble with bearings or cylinder walls cutting. Grooved cylinder walls permit the hot gases to blow by the piston rings and overheating results. Sometimes the internal oil conduit or pipe running from the back end of the crank case to the front may be partially clogged with gummed oil or

112

wax particles from the oil. This pipe should be cleaned out by a blast of compressed air at high pressure from time to time. It can be reached from the top of gear case when the cover casting is removed. The oil collecting wells that direct the lubricant to the main bearings may fill with dirt or other matter, such as gummed oil, which prevents proper quantities from reaching the bearings. This condition is apt to occur if the crank case is not drained out and cleaned from time to time as recommended.

Deposits in Radiator and Piping. — The form of radiator most generally used at the present time has a number of very narrow passages through which the water must pass in going from the upper compartment, into which it is discharged after leaving the motor cylinders to the lower compartment, where it collects after being cooled and from which it is drawn by the circulating pump. The water used in some localities for cooling contains much matter, either in suspension or solution, which will form scale or a powdery deposit in the radiator tubes. It does not take much scale to seriously reduce the ratio of heat conduction between the heated water inside of the tube and the cooling air currents which are circulated about their exterior. As Ford cylinders are of cast iron, a certain amount of rust will be present in the water jacket, and this also may get into the radiator piping.

If an anti-freezing solution using some salt as a basis, such as calcium chloride, is employed, after this has been circulated through the radiator and piping for a time it may deposit solid matter in the form of crystals. Antifreezing solutions that include glycerine may have a chemical action due to the acid sometimes found in the cheaper commercial grades of glycerine employed for this purpose. This chemical action results in the deterioration of the water jacket walls, and also contributes to the rust deposit.

For cleaning out water spaces of a radiator a solution of potash or caustic soda may be used. This will cut the rust and some forms of scale, and will dissolve them or loosen them sufficiently so the deposits may be thoroughly flushed out with water or steam under pressure. The solution will work more rapidly if it is brought to the boiling point before placing it in the radiator. The solution is also valuable in removing rust from the water jacket interior. The best action is obtained if the caustic soda solution has a strength of between 15 and 22 per cent.

In order to apply this method, the whole of the water in the circulating system is drained off and measured. Then a solution is made by dissolving 2½ lbs. of solid caustic soda so that it makes one gallon of solution, enough of the solution being made to fill the entire cooling system.

Considerable heat is generated when the soda is dissolved, and frequent stirring is necessary, unless the soda is hung in an iron basket just under the surface of the liquid. When the liquid has cooled it is introduced into the circulating system until the latter is entirely filled. The soda is allowed to remain in the system all night and is run out in the morning. It must be borne in mind that caustic soda will corrode aluminum in other cars and must not be used if the system should have an aluminum pump housing. After running out the soda a hose pipe and water supply should be connected to the system and a good stream of water driven through at fair pressure for some time. It will not harm Ford parts.

Incrustation is most commonly caused by carbonate of lime, which is held in solution in some waters as a bicarbonate; therefore, when the water is heated the carbonic acid is driven off and the carbonate is precipitated in the form of a muddy deposit, which hardens in the presence of heat into a non-conducting scale in those portions of the water jacket where the heat is greatest and which remains in the form of a powdery deposit in the radiator tubes, where the heat is not great enough to harden the sediment. Sometimes the deposit is sulphate of lime, this also being found in the water available in some localities. The reason that water contains so many impurities is because it is one of the best-known solvents. Pure water is never found 4n nature and can only be obtained by a process of distillation. The purest natural water is rain, and if this is collected before it touches the earth it contains only such impurities as may be in the air.

Method of Pan Belt Adjustment. — If the motor heats up when the engine is running and the car standing still, it is necessary to inspect the fan driving means to make sure that this is functioning properly and that the fan is turning all the time the engine is running and at the proper speed. Ford cooling fans are flat belt driven and are mounted on a simple form of bracket that will permit of maintaining the fan driving belt at the proper tension to insure positive rotation of the fan blades.

The fan belt tension may be easily adjusted by loosening a lock nut on a cap screw which bears against the lever carrying the fan hub and tightening down on the belt tension regulating screw until the belt is tight enough so that a decided resistance will be felt due to fan belt friction when the fan is turned by hand. Care should be taken not to tighten the fan belt too much, however, as this will result in rapid wear of the belt and cause it to stretch as well as imposing some strain on the fan hub bearing. The adjusting screw in the fan bracket may be easily reached by removing the small brass cap which covers the breather pipe at the front end of the engine. If the water in the radiator boils easily and the trouble is not due to heavy work, it is usually because the fan belt is loose.

Among some of the troubles that will cause overheating are: too much driving on low speed; not enough lubricating oil; carbonized cylinders or combustion chambers; spark retarded too far; clogged muffler; insufficient lift to exhaust valves, poor carburettor adjustment, clogged radiator tubes, fan not working properly on account of broken or slipping belt, and leaky piston rings. In examining the cooling system, one should look for a leaky radiator, jammed or broken radiator pipes, bad connections in the rubber hose used to join radiator and water jacket, lack of water and sediment in radiator or water jacket. The Ford Company does not advise the use of nostrums such as com meal, bran or other substances put into the radiator for the purpose of stopping leaks, as these are bound to hinder proper circulation. The interior walls of the rubber hose may decompose and particles of rubber hang down, partially constricting the hose bore and hindering the passage of water. As new hose connections are inexpensive and easily applied, defective members should be replaced at once.

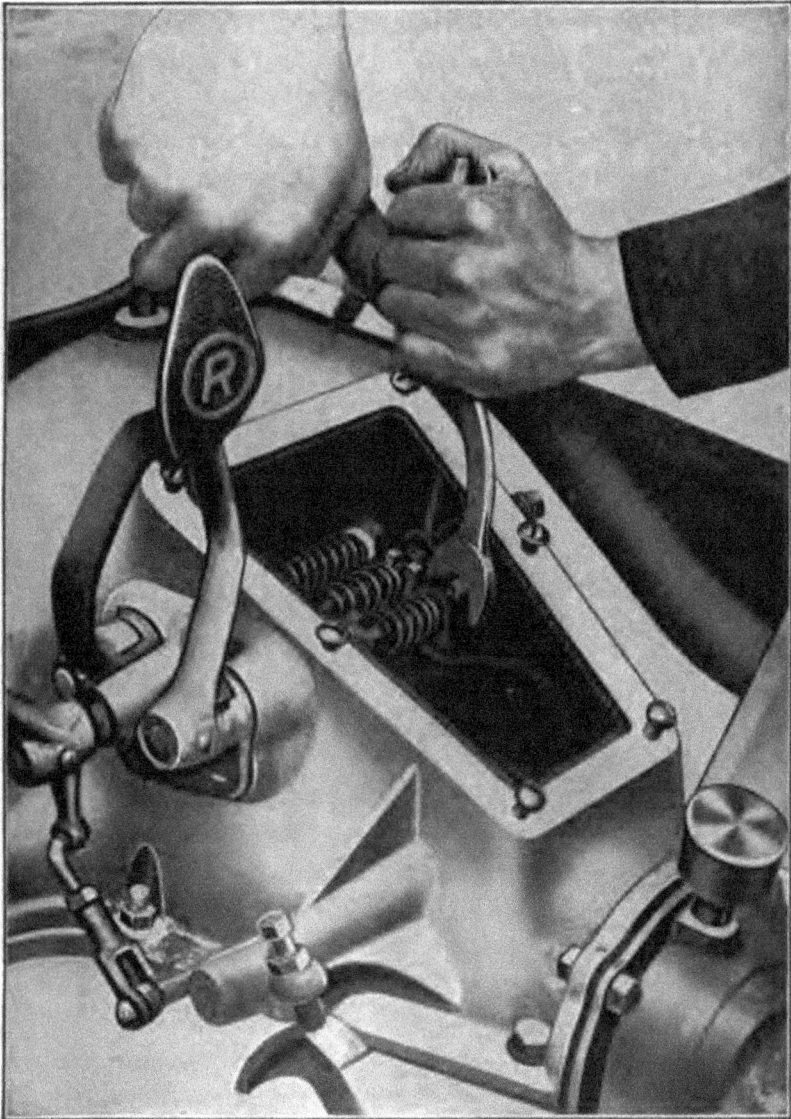

Fig. 60.—Method of Adjusting Transmission Brake Band.

Adjusting Transmission. — If any difficulty is experienced in climbing hills on the high speed and the engine seems to have adequate power it is because the high speed clutch is slipping. If the engine races when the slow speed pedal is pushed up as tight as it will go when climbing steep hills it is because the slow speed band in the transmission needs adjusting. Should the car show a tendency to creep forward when the crank is turned for starting the motor, this indicates that the clutch lever screw which bears on the clutch lever cam has worn and requires an extra turn down to hold the clutch in neutral position. This screw and the cam on which it works are clearly shown in the lower portion of Fig. 60. As will be observed, the screw passes through a boss on the end of a simple lever, the other end of which is joined to the clutch actuating pedal by a simple rod and yoke connection. The lock nut must be released before attempt is made to screw down the clutch lever screw, and care must be taken not to screw that down any more than absolutely necessary or the high speed clutch will slip.

Fig. 61.—How to Tighten Slipping High Speed Clutch.

To make adjustments to the reverse or transmission brake band, or to the high speed clutch, the plate on the transmission cover must be removed, this exposing the interior of the transmission, as shown at Fig. 60. It will be necessary to remove the floor boards of the car to get at this portion of the power plant. In order to simplify the point involved, the method of adjusting a high speed clutch is shown at Fig. 61, in which the top half of the transmission case has been entirely removed. It will be noticed that there are three clutch fingers spaced around the front end of the transmission drum. These

116

have set screws, through which the pressure of the spring is translated to push members bearing against the clutch disc assembly. These set screws are prevented from turning by split pin locks. If the high speed clutch is slipping, the split pin on the clutch finger that locks the set screw should be removed and the set screw given one-half or one complete turn to the right with a screw driver. Each of the other set screws should be turned exactly the same amount, and care should be taken to replace the locking split pin.

Fig. 62.—Testing for Front Wheel Looseness.

The low speed band may be tightened without removing the transmission cover plate, as an adjusting screw is provided outside of the transmission case and on the right hand side. If the slow speed band does not hold positively the lock nut on this adjusting screw may be loosened and the screw turned toward the right until appreciable resistance is felt when the clutch pedal is pushed way forward. This adjusting screw is shown in the space between the transmission case and the exhaust pipe in Fig. 60. The foot brake and reverse bands cannot be adjusted without removing the transmission case cover plate, as shown at Fig. 60. Adjusting nuts will be found that may be readily turned with an S wrench, as indicated, until the proper degree of band tension has been obtained by bringing the ends closer together. Care should be taken not to tighten the bands too much, as they may drag on the transmission drum assembly when the car is in the high speed, act as a brake and tend to overheat the motor. The foot brake should be adjusted tightly enough, however, so that medium pressure on the foot pedal will stop the car immediately or slide the rear wheels in case of emergency. The reverse band should be tightened in the same way as the brake band is. If the friction linings of the bands are worn to such an extent that adjusting the brake does not seem to improve matters, they should be removed and relined with new friction material so they will engage smoothly without causing a jerky movement of the car, as is unavoidable with slipping bands. These linings are inexpensive and with ordinary good judgment a reverse and slow speed band should last at least 10,000 miles and the foot brake lining over 5,000 miles. Instructions for removing the brake bands are given in the nest chapter, the purpose of this paragraph being merely to explain the adjustments possible on the road.

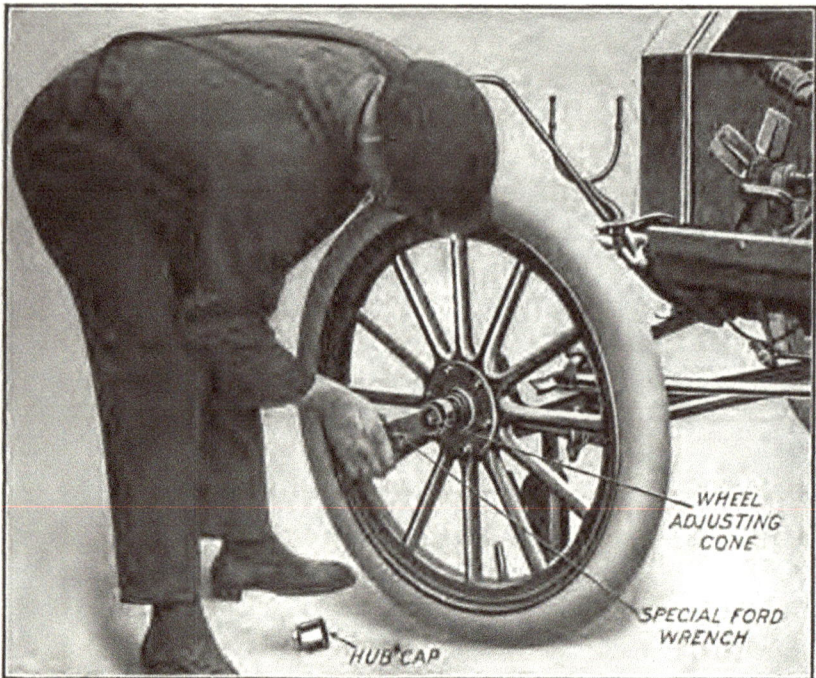

Fig. 63.—Showing Use of Special Ford Wrench in Adjusting Front Wheel Bearings.

Adjusting Loose Front Wheels. — At regular intervals while the car is in service it is important to examine the front wheels to make sure that these are retained properly by the front hub ball bearings. The method of testing the wheel for looseness is clearly shown at Fig. 62. The front axle is raised

118

from the ground by jacks and the wheel is grasped at top and bottom by the operator, who attempts to push the top of the wheel in and pull the bottom out, or the reverse of this operation. If the spindle is tight in the front axle yoke and there is considerable looseness in the wheel hub, the hub cap should be removed and the bearings carefully examined. The cone should be removed from the spindle by pulling out the split pin that locks the castle nut, remove that member, as well as unscrewing the cone. After the cone is removed from the threaded end of the spindle the wheel may be readily withdrawn. The cones on both inner and outer parts of the spindle should be carefully examined to make sure that they are not pitted or scored, and the balls and races in the wheels should also be examined for defects. If these parts are found in proper condition the proper wheel adjustment may be secured after the wheel is replaced by tightening on the wheel adjusting cone up to the point where the wheel will turn freely and have no appreciable end play. After the proper adjustment of the cone has been secured, the lock nut and split pin should be replaced. The hub cap should be filled with grease and replaced on the wheel hub. The interior construction of the Ford front wheel may be clearly understood by referring back to Fig. 32, which shows a sectional view of the Ford wheel hub and ball bearings in place.

Fig. 64.—Method of Testing Rear Wheel Brakes, Also Wear in Axle Bearings.

What to Do When Rear Brakes Do Not Hold. — After the car has been in use for a time trouble may be experienced in holding it on a grade by locking the emergency brake hand lever. This is because the rear brake shoes have worn sufficiently so that they do not grip the brake drum interior in a positive manner. In some cases this wear may be taken up by shortening the rods running from the hand lever cross shaft to the brake cam levers. This may be done by the adjustment provided in threaded rod ends.

The method of testing the rear wheel brake is shown at Fig. 64. Two jacks are utilized to raise the rear axle to have both wheels clear of the ground. The hand brake lever is then applied and endeavor is made to move the wheel by grasping opposite spokes as shown, pushing down on one and pulling up on the other. If the wheel can be moved it is a sign that the brake shoes do not hold. It is possible to compensate for some wear of the brake shoes by shortening the rods, but it is important when this is done to have these of the same length in order that one brake will not take hold quicker than the other.

Fig. 65.—How to Use Wheel Puller for Removing Wheel From Taper Axle End.

To examine the brakes it is necessary to remove the wheel, which is done by taking off the retention or clamping nut from the end of the axle and then applying a special wheel puller in place of the hub cap, as shown at Fig. 65. This wheel puller consists of a casting threaded to fit the wheel hub at its lower end and carrying a set screw at the closed end. After the wheel puller body is properly screwed onto the hub and a clamp screw tightened to make

120

Fig. 66.—Wheel Removed to Show Internal Expanding Brake Construction.

sure that the puller body will not come off and strip the threads, the wheel may be forced off of the taper by screwing in on the set screw with a large monkey wrench furnished with the Ford tool outfit, at the same time keeping the wheel puller body from turning with the hub cap wrench. If any difficulty is experienced in removing the wheel after a certain amount of tension is obtained by screwing in the set screw, a few sharp blows with a wooden mallet or lead hammer on the head of the set screw and applied in the direction of the axle will loosen the wheel. After the wheel is removed the brake shoes may be readily examined as they are exposed, as shown at Fig. 66. It sometimes happens that the slipping is caused by oil deposits rather than wear. In this case the slipping is easily remedied by cleaning out the interior of the brake drum and the faces of the brake shoes thoroughly with gasoline. If the shoes show signs of wear, which can be ascertained by noting if some portions are worn thinner than others, new brake shoes should be obtained and installed in place of the worn members. All that holds the brake shoe assembly is one retention bolt carried by the plate at the end of the axle. "When this bolt is removed the brake shoe assembly can be pulled away from the axle and the expander cam. It is much cheaper to replace worn shoes with new ones, which are inexpensive, than to attempt to get further service from the worn brake shoes by building upon the faces, which are in contact with the expander cam, as sometimes advised, or by attempting to apply a facing

121

of sheet metal to the brake shoes. When new cast iron shoes are installed the wheel should be pushed on the axle taper and turned around with the hand brake lever in neutral position to make sure that there are no rough spots that will bear against the brake drums and cause friction, make a squeaking sound, and produce heating of the brakes when the car is running forward on the high speed. If any rough spots are found they should be smoothed off with a file. When reassembling the wheel care should be taken that the key used to drive the wheel is in place and that the hub is bedded tightly on the taper by the retaining nut, which must be screwed on tightly and locked with a split pin to prevent it backing off. If the brake shoes have been replaced after the actuating rods have been shortened several times it should be borne in mind that these rods must be lengthened again to normal length when new brake shoes are fitted.

Chapter Five - Overhauling and Repairing Mechanism

After any automobile has been in use for a time, even if it has always been driven very carefully and oiled regularly according to the makers' instructions a certain amount of depreciation will exist in the mechanism and the car can only be put in good condition again by a thorough overhauling process. This is work for the skilled mechanic or owner with mechanical inclinations and a knowledge of the use of tools rather than the average owner or driver. The writer will endeavor to give sufficiently clear instructions however, so that the Ford owner who knows how to handle common tools and who has a general knowledge of common mechanical processes will be able to make many of the repairs mentioned in this chapter. To further facilitate an understanding of the descriptive matter photographs have been taken of numerous parts of the Ford chassis mechanism in various stages of assembly and careful study of these will give even the non-mechanical owner some information that may be of value.

When a car needs overhauling it gives notice that this attention is needed by readily recognized symptoms. For example, the engine will lack its customary power and speed and hills that could formerly be negotiated on the high gear can be climbed only by the use of the low speed ratio. The engine will not accelerate properly nor run regularly and it will be noisy in action. The gears will grind all the time the machine is in use and the many minor chassis parts that have worn will rattle whenever the car is operated on anything but the best of highways. The brake bands in the transmission will have reached the limit of their adjustment, the brakes in the rear wheels will no longer hold positively when the hand lever is brought completely back in its travel. The car will not be as responsive to steering and considerable lost motion will be evidenced at the steering wheel which must be turned an appreciable amount in order to start to move the wheels. It will be apparent that while the car is still in an operative condition that it is not working efficiently. It is not desirable to run it when in need of mechanical repairs because the faulty parts are wearing more rapidly after initial depreciation sets in. The first point to consider will be the restoration of the mechanism of the power plant, after which the repair of the transmission and chassis parts will be discussed.

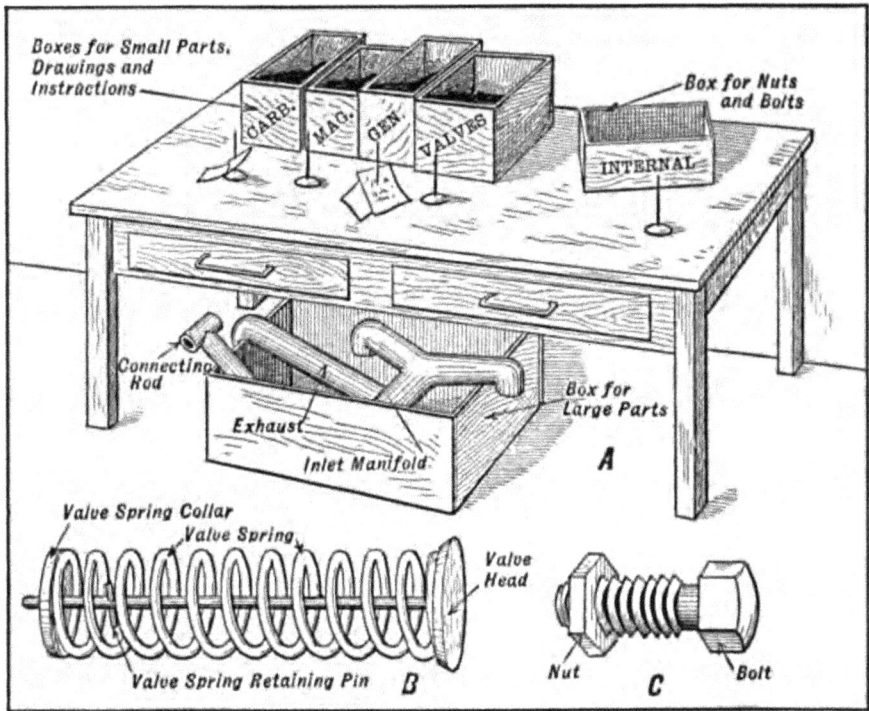

Fig. 67.—Showing Method of Providing Boxes for Keeping Parts Together in Overhauling Cars Systematically.

Faults in Power Plant and Symptoms. — There are a number of unmistakable symptoms that indicate depreciation of power plant components. The most common of these is lost power, then comes noisy operation. If an engine loses power it is not always a sign that it is in need of a complete overhauling, as the trouble may be due to local causes which can be readily remedied without taking the engine out of the frame. If the lost power is accompanied by noisy action it is safe to consider that the trouble is due to wear of the motor parts. If an engine is cranked over by hand and the compression is poor, one may assume that either the valves need grinding or that the piston rings are worn or the cylinder walls scored. Back firing in the carburettor or inlet pipe, if it is not due to lean mixture, would be caused by

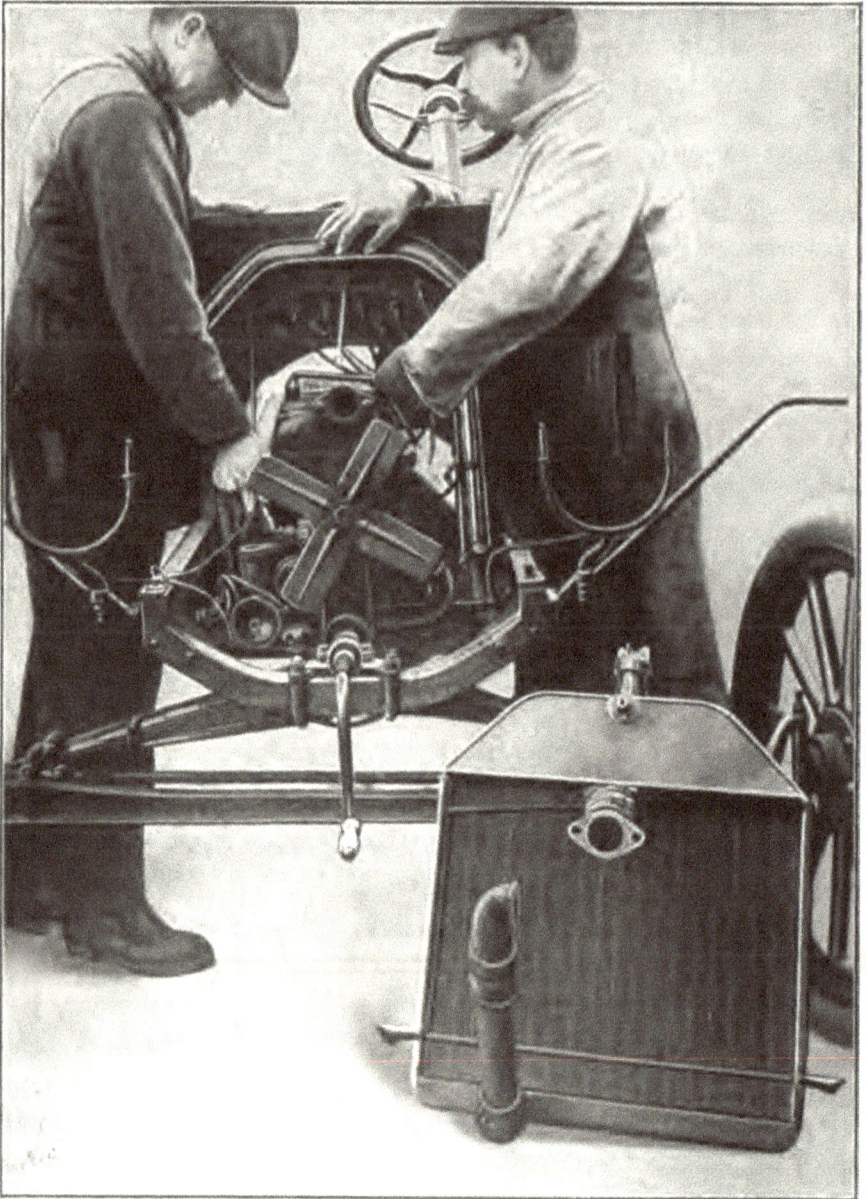

Fig. 68.—The First Step in Removing the Ford Engine From the Chassis Is to Take the Radiator From the Front of the Frame.

leaky valves, the inlet valve not closing properly, defective inlet valve springs, or incorrect valve timing. If the engine has no power and the crank case gets hot, this is a sure indication of burned gas leakage past the piston rings which have become worn or broken. Noisy action in the engine, if not due to carbon deposits or insufficient lubrication would be the result of worn pistons and cylinder walls, worn wrist pin bushings in the piston bosses, loose connecting rod bearings at the crank shaft, loose main bearings, flywheel loose on crank shaft, camshaft driving gear loose on its retaining key, too much space between valve operating plungers and valve stem, or loose camshaft bearings. If the loss of power is accompanied by overheating and the cooling and lubricating systems are functioning properly, trouble is probably due to insufficient lift of the exhaust valves or the muffler may be clogged with carbon deposits, these preventing a free exhaust. The best way to overhaul the engine is to take it out of the frame and place it on a bench where it can be reached handily and the parts inspected more easily. Before describing the method of taking down the Ford motor, it may be well to consider the subject of systematic procedure in overhauling, as this is a point that is often neglected by the professional repairmen as well as the amateur mechanic

System in Overhauling. — Not a few motor car owners are backward in overhauling the motor, having in mind continually the fear that it will be a difficult matter to get the parts together again properly. While it is a comparatively simple procedure to remove parts it is quite the opposite in assembling so that the whole will be as it was originally. If a system is followed throughout there will be little if any trouble in resetting the motor parts or in fact any parts and for the benefit of the inexperienced owner a few hints on the subject that have appeared in Motor Age are given here. Before anything is done to the motor the workbench should be put in shape. A number of boxes of varying sizes should be placed along the back of the bench as shown in Fig. 67. Each box is designed to receive the small parts of some unit such as the engine cylinder parts, valve system, etc., and each box should be properly marked with the name of the parts it is to contain. In front of each box a spindle file should be placed. These files are to be used for securing data sheets upon which are written notations concerning the various adjustments. Suppose difficulty should arise in removing a part, the first question to enter one's mind is ^^will it not be more difficult to return it to position? ' ' Such a condition calls for a notation on a slip of paper and the exact location of all the parts drawn if necessary. In other words the files are used for memory so that the operator will know just how every part in each box was removed and the position it occupied with reference to some other object.

Adjustments often are required to be altered so that memory slips again are valuable in that the original adjustment can be jotted down. Aside from the small boxes on top of the bench a large box should be placed underneath. This is to care for large parts such as intake and exhaust manifolds, connecting rods, the fan, the carburettor and other parts of the assembly. In other words, the large box receives parts of which there is no doubt as to location. System in removing parts will eventually lead to a saving of time when assembly is necessary. For example when the valve system is removed, the valves, the springs, spring seats and the holding pins should be kept together as shown in illustration Fig. 67, B. In removing bolts, the lock washer, the nut and every part belonging to the bolt should be kept together as at Fig. 67, C. Slip the washer over the bolt and screw the nut in place and then throw the whole into the proper box. When the wrist pins are removed the clamp screw for holding the pin in place should be left in the hole in the connecting rod boss. When the manifolds are removed, instead of throwing the holding nuts into one box and the manifold in another, screw the nuts back onto the studs in the cylinder. The object of all this immediate replacement is to keep the parts as near to their proper places as possible so that when assembly begins the operator will not be in doubt as to where to look for a certain rod bolt or other part.

How To Take Down Motor. — The first step in removing the Ford engine from the frame is to drain the water from the radiator by opening the pet-cock at the bottom and while the water is draining out to disconnect the spark plug wires at the top of the motor. After the water is drained out the water connections are unfastened from the top of the motor and the side of the water jacket leaving the hose and water pipes attached to the radiator as clearly shown in Fig. 68. The next step is to disconnect the radiator brace rod which holds it to the dash by unscrewing it out of the lug at the top of the radiator. In order to unscrew this rod it will be necessary to loosen the check nut on the end of the rod that bears against the dash. The two nuts that fasten the radiator to the frame are removed and the radiator may be taken off and placed to one side of the way, care being taken to place it in a position where it will not be injured by tools or other parts falling on the cooling tubes.

Next disconnect the dash at the two supporting brackets which rest on the frame and detach the steering post guide bracket carried at the left hand side of the frame. The timer case or cover is loosened by releasing the retaining clamp spring as shown In illustration and the circular loom containing the four timer wires running to the coil is released from the small clamp that holds it to the frame member in some cases. This makes it possible to remove the steering post, dash, and all the wiring as one assembly. Before taking off the steering post and dash assembly it is necessary to disconnect the carburettor control, mixture adjustment and timer cover advance rods. Then the two bolts of the cap holding the ball at the apex of the triangular front radius rod member are unscrewed from the socket underneath the crank case. The next step is to free the rear of the engine from the rear axle unit, which is done by removing four bolts at the universal joint ball housing. The next step is to turn off the gasoline at the tank, disconnect the fuel supply pipe from the carburettor, and remove the pressed steel pans joining the cylinder casting to the frame on each side. The exhaust manifold is uncoupled from the exhaust pipe by unscrewing the large brass packing nut. Next release the cap of the trunnion bearing which supports the front end of the crank case on the front frame cross member which is done by unscrewing the two cap screws that hold the cap in place. The crank case supporting arms are attached to the frame side members by two bolts in each arm. These bolts should be re-moved which breaks the last connection holding the motor to the frame.

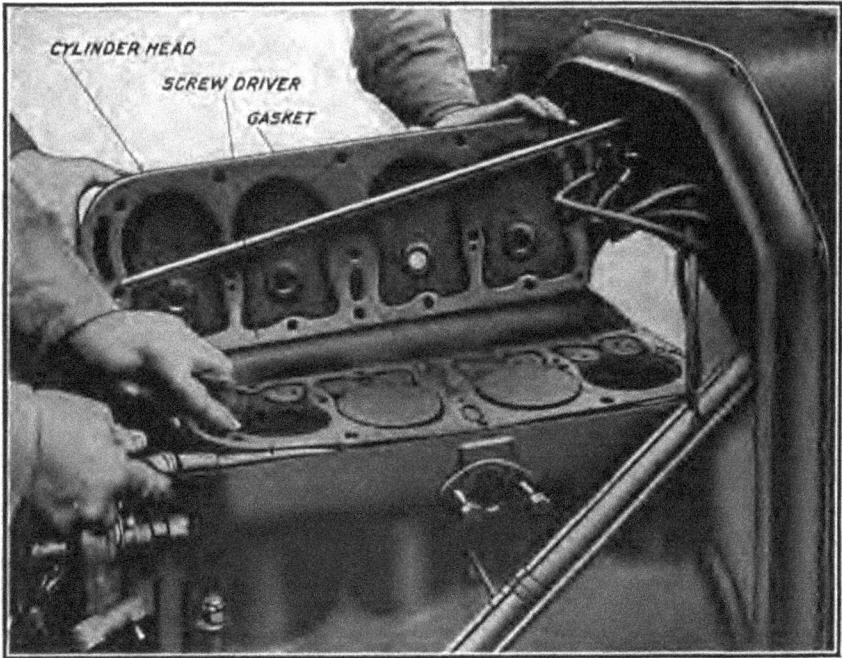

Fig. 69.—Showing Method of Removing Copper Asbestos Gasket Used Between Cylinder Head and Cylinder Block Casting of the Ford Engine.

A stout rope is passed through the opening between the two middle cylinders and is tied with some form of positive holding but quick releasing knot. A piece of 2" x 4" studding or a substantial iron pipe about 10 feet long is passed through the rope. The engine may be lifted out of the frame by three men, one at each end of the beam or pipe while the third man takes hold of the starting crank handle to steady the power plant while it is being lifted. If a chain falls or portable crane is available one man may easily handle the power plant.

After the engine is removed the first step in taking it down is to release the fifteen cap screws which hold the cylinder head in place and take the head easting off. Next the bottom plate on the crank case is removed and all the connecting rods loosened up which permits of pushing the pistons out through the top of the cylinder if desired. The cooling fan and the fan belt are removed and placed with the radiator where they will be out of the way. At Fig. 68 the dash assembly is just loose enough to be raised from the frame when necessary and one of the workmen is loosening the cylinder head re-tention bolts before the engine is lifted out of the frame on account of the superior purchase or grip afforded when the engine is in place. The workman

127

at the right is loosening the fan bracket adjusting bolt in order to take off the fan.

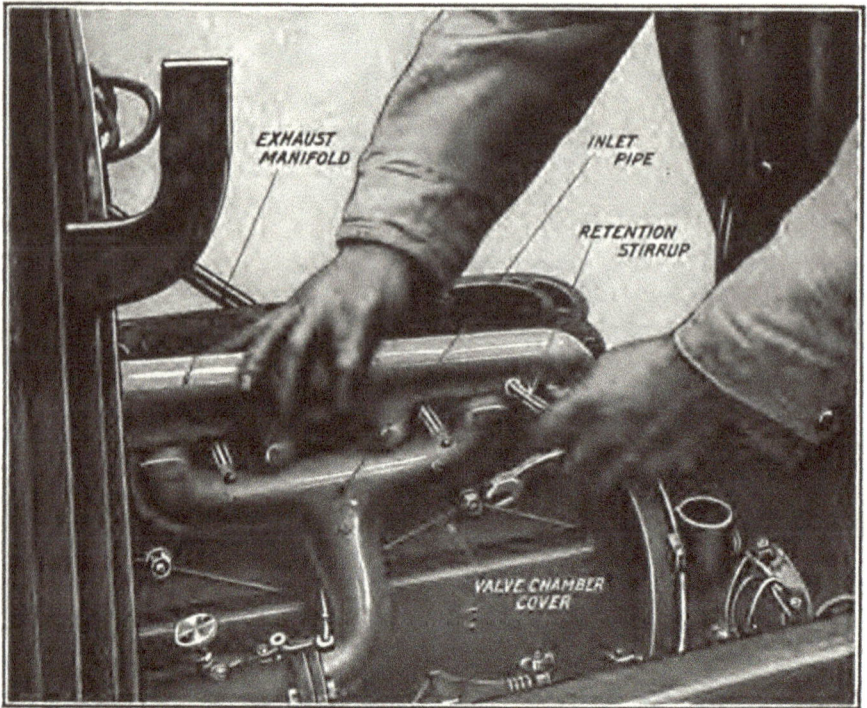

Fig. 70.—Method of Removing Inlet and Exhaust Manifold.

After the cylinder head is removed the next operation is to carefully take off the gasket which will be found on top of the cylinder casting. This is done by prying it away from the cylinder block gradually with a screw driver or other similar implement. The next step is to remove the inlet and exhaust manifolds which is done by unloosening the stirrups or clamps holding these in place as shown at Fig. 70. This exposes the valve chamber covers which may be removed as outlined at Fig. 71. The method of taking the piston out through the top of the cylinder head after the connecting rod cap has been removed from the crank shaft is clearly shown at Fig. 72. The top half of the transmission case is then lifted off, which can only be done after the brake band adjustments have been loosened up and a number of retention bolts removed. The final operation is to take off the pressed steel lower crank case member which also forms the lower portion of the transmission case compartment. The appearance of the bottom of the engine after this is done is clearly shown at Fig. 73.

It is important to keep all parts that come from any special member together and mark them so that they may be readily identified. Many experienced repairmen have the habit of throwing all parts in a common box and then picking them out as needed. While the expert has no difficulty in distinguishing the pieces much time is lost in looking for the various bolts and nuts, to say the least, while if the novice follows this practice he will be hopelessly confused and will have difficulty in identifying the various pieces. If the pistons and connecting rods are not already marked they should be plainly stamped with steel numbers or letters or with a series of prick punch marks to make sure that they will be replaced in the same cylinder from which they were removed.

Fig. 71.—Valve Chamber Cover Plates Must Be Removed to Gain Access to the Valve Springs.

Carbon Deposits and Their Removal. — Mention has been previously made that carbon deposits in the combustion chamber are not desirable because they are apt to produce overheating and noisy operation and result in diminution of power. The knock produced by carbon is a clear hollow sound, generally evidenced when climbing grades on the high speed, especially after the engine has become heated. Carbon deposits are also indicated by a sharp knock noticed whenever the engine is speeded up by opening the throttle. The knock produced by having the spark too far advanced is duller than that which is caused by carbon. A loose connecting rod knock sounds like the tapping of steel with a small hammer and is most easily distinguished when a car is allowed to coast down grade or upon suddenly slowing up the car from speeds of 25 to 30 miles an hour by closing the throttle. Looseness in the crank shaft main bearings produces a knock which can be best distinguished when the car is going up hill or under road conditions where the engine is working hard. The knocking sound produced by a loose piston is heard when the throttle is suddenly opened and is almost exactly the same as a carbon knock.

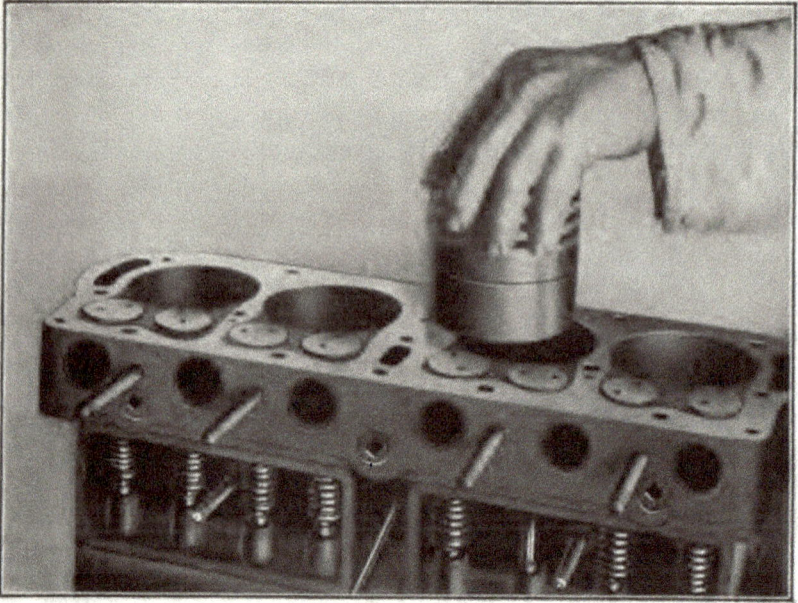

Fig. 72.—How Ford Piston May Be Withdrawn Through Top of Motor.

It is not necessary to take the engine entirely out of the frame only to remove carbon or grind valves, though if these two processes are to be done at the same time the motor is overhauled it will be just as well to do all that is necessary to the engine after it has been taken out of the frame. If one desires merely to remove carbon, the first step is to drain the water out of the radiator and to disconnect the water connection bolts at the cylinder head.

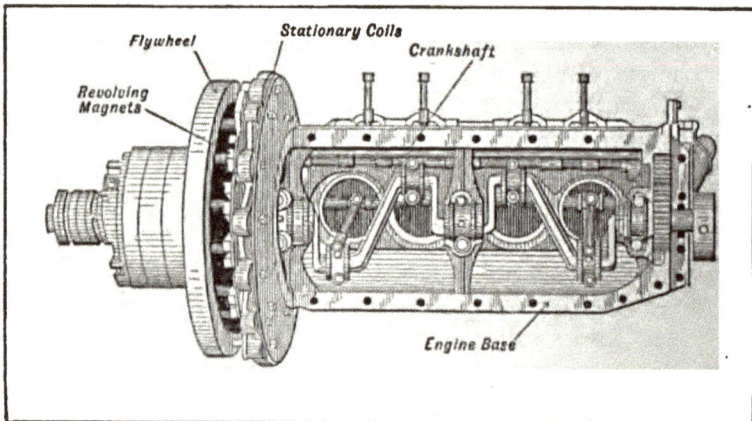

Fig. 73.—Bottom View of Ford Engine With Pressed Steel Lower Crank Case Member Removed.

The wires should be disconnected from the spark plugs in the cylinder head and these should be removed in order that they may be cleaned and placed out of harm's way. If the spark plugs are left in the cylinder and a wrench is applied to a cylinder head bolt and slips off, one is very likely to break the insulation of the spark plug which is of porcelain, a very brittle material. It is not necessary to remove the radiator when the cylinder head is taken off for scraping carbon from the piston tops or grinding valves.

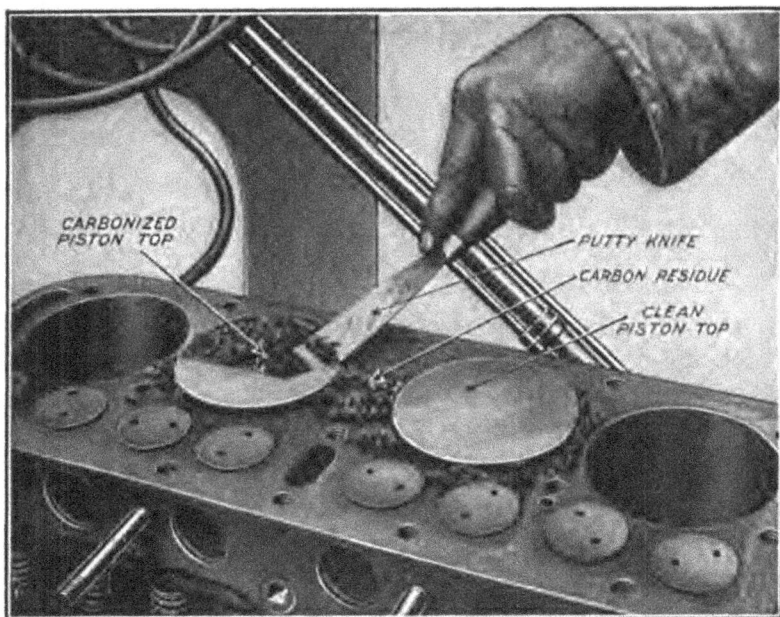

Fig. 74.—Method of Removing Carbon Deposits from Piston Top and Cylinder Block.

After the cylinder head has been removed it should be inverted on the work bench and the spark plug holes plugged up by screwing in the plugs. The combustion chambers are then filled with kerosene which is allowed to soak in to soften the carbon deposits while those on the piston top are scraped off with a putty knife, as shown at Fig. 74. Care should be taken to prevent the particles of carbon from getting into the cylinders, bolt holes or water jacket openings. It is evident that the gasket must be removed from the cylinder block as shown at Fig. 69 before the carbon is scraped away. If the carbon removal process is to be followed by valve grinding it will be well to take off the inlet and exhaust manifolds as shown at Fig. 70 and the valve chamber cover plates as outlined at Fig. 71. Care should be taken in replacing the cylinder head gaskets to have the pistons in the cylinders No. 1 and No. 4 at top center, using these as guide members to locate the gasket in position and to locate the cylinder bead in place.

Be sure to draw the cylinder head retaining bolts down evenly, turning down each one only a few turns at a time. Do not tighten the bolts at one end before the other, but all should be given the final turn to bed them down at practically the same time. The best way is to turn those in the center of the casting down so they bed loosely first, then to tighten one at the front and one at the rear. After this all of the bolts may be screwed down tightly. After the cylinder head is replaced the engine crank shaft should be turned over with the starting handle so as to be sure that the gasket does not project over the cylinder bore, which would mean that the piston would hit it when it came to the top of the stroke. Before replacing the cylinder head the combustion chamber should be thoroughly cleaned out and all carbon removed by scraping. If the kerosene has been used as recommended, in many cases the carbon deposits may be softened sufficiently so they may be wiped out with a rag.

How to Repair Cracked Water Jacket. — The water jacket of the Ford engine cylinder will sometimes become cracked due to freezing of the cooling water or perhaps as a result of a sand or blow hole which opens up from vibration after the cylinder has been used awhile At the present time the usual practice in repairing cylinders is to fill the depression or crack with iron by the autogenous welding process, although various iron cements may be used for that purpose if the fracture is not serious. A mechanical repair is always possible, i.e., a metal patch can be applied to cover the crack and held in place against a gasket interposed between the plate and the cylinder jacket by small machine screws tapped into the iron.

If the crack is of some length it may be repaired by the following method: On the line of the fracture, drill and tap for a 3/8" threaded copper rod. This rod is screwed in firmly to a depth about equal to the thickness of the metal of the water jacket. Cut off the copper rod with a hacksaw, allowing it to project about 1/32"; then drill succeeding holes, each hole being drilled partly into the previously inserted copper plug, so that when all of the plugs are placed in the cylinder casting, they form a continuous band of copper along the line of the fracture. The copper plugs should be peened down and trimmed off flush. The only possible chance for leakage, after having repaired the crack in this manner, is for the water to follow the joint between the metal of the jacket and the copper plugs, but as the copper rods are threaded into the casting, it is not likely to occur. Should leakage take place, a little extra peening will suffice to prevent it.

Still another method involves fusing copper filings or granulated brass spelter into the crack. This has the advantage of not requiring the removal of the part to be repaired. Drill and tap a small hole at each end of the crack to prevent further extension of the weakness, and screw in an iron stud. Next clean the outside and inside of the fracture very thoroughly, using a scraper and gasoline. File up some soft copper or brass spelter, mix with borax and fill the crack, heaping the filings over it. Then take a powerful blow lamp or a torch and direct the flame on the copper. By this method a fair amount of metal can be worked into the opening. After cooling, the studs are cut off flush and the copper filed smooth. It is said that the repair will endure indefinitely.

In many cases where the crack is not serious it may be closed by making a rust joint. The first step is to drill a very small hole at each end of the crack to prevent it from spreading and to drive in or screw in a metal plug in each hole. The crack is then filled up with a paste made of 66% iron filings or iron dust and 33% sal ammoniac in the pulverized form, with just enough water to make the mixture of proper consistency to be pressed into the crack easily. The action of the sal ammoniac is to rapidly oxidize the fine iron filings, producing rust which joins the various iron particles together and effectively seals the opening when it has properly hardened. As a number of prepared cements for use with cast iron may be purchased at low cost it is often cheaper to buy the cement than to attempt to make it.

Another method sometimes employed is to clean out the interior of the water jacket thoroughly and put in a solution of sulphate of copper or bluestone, allowing this to leak through the crack, if it will. Care is taken to remove any traces of grease that may remain in the crack; this may be washed out by a boiling hot solution of potash or caustic soda. As the copper sulphate solution leaks out, it deposits a thin copper film, and if the crack is such that it permits only a slow leak, the defective point will be sealed overnight with a deposit of pure copper.

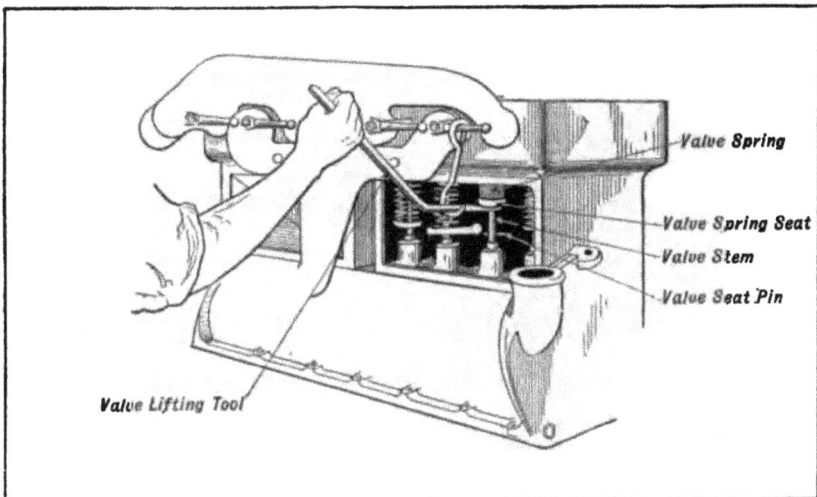

Fig. 75.—Method of Compressing Valve Spring in Order to Remove Valve Seat Pin by Using Ford Valve Spring Lifter.

Reseating and Truing Valves. — Much has been said relative to valve grinding and despite the mass of information given in the trade prints it is rather amusing to watch the average repair man or the motorist who prides himself on maintaining his own car performing this essential operation. The common mistakes are attempting to seat a badly grooved or pitted valve on an equally bad seat, which is an almost hopeless job, and of using coarse emery and bearing down with all one's weight on the grinding tool with the hope of quickly wearing away the rough surfaces. The use of improper abrasive material is a fertile cause of failure to obtain a satisfactory seating.

Valve grinding is not a difficult operation if certain precautions are taken before undertaking the work. The most important of these is to ascertain if the valve head or seat is badly scored or pitted. If such is found to be the case no ordinary amount of grinding will serve to restore the surfaces. In this event the best thing to do is to remove the valve from its seating and to smooth down both the valve head and the seat in the cylinder before attempt is made to fit them together by grinding. Another important precaution is to make sure that the valve stem is straight, and that the head is not warped out of shape.

A number of simple tools are available at the present time for truing valve seats, one of these being outlined in use at Fig. 77. It is a simple reamer having cutters conforming to the valve seat angle and a T-handle by which it may be turned.

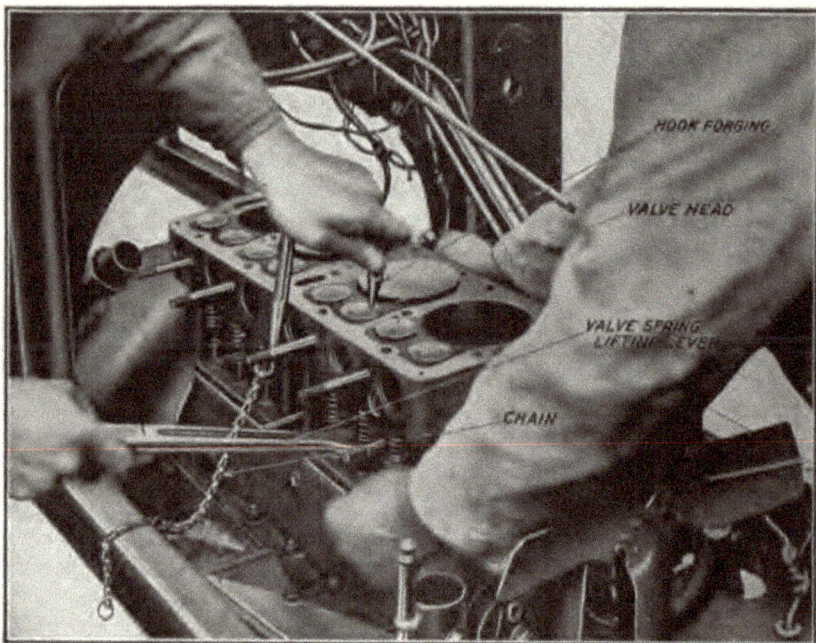

Fig. 76.—Special Tool for Raising Valve Springs When Inlet and Exhaust Manifolds are Taken Off of the Cylinder Block.

Method of Valve Grinding. — The process of seating valves by grinding is not a difficult one, requiring patience more than mechanical skill. The first step is to remove the valves, which may be accomplished as shown at Fig. 75, using the regular Ford tool, or at Fig. 76, using another type of valve spring compressor. Before the valves can be removed, the spring that keeps them seated must be raised enough to permit the removal of the valve seat pin shown at Fig. 75. After the pin is pulled out of the valve stem, the valve may be pushed up from its seat and removed for examination. The writer believes that it is better to remove the manifolds before grinding the valves, so the tool shown at Fig. 76 is best to raise the valve spring as it does not depend upon a manifold retaining stirrup as an anchorage. Another advantage is that it keeps the valve head against the seat while the spring is compressed.

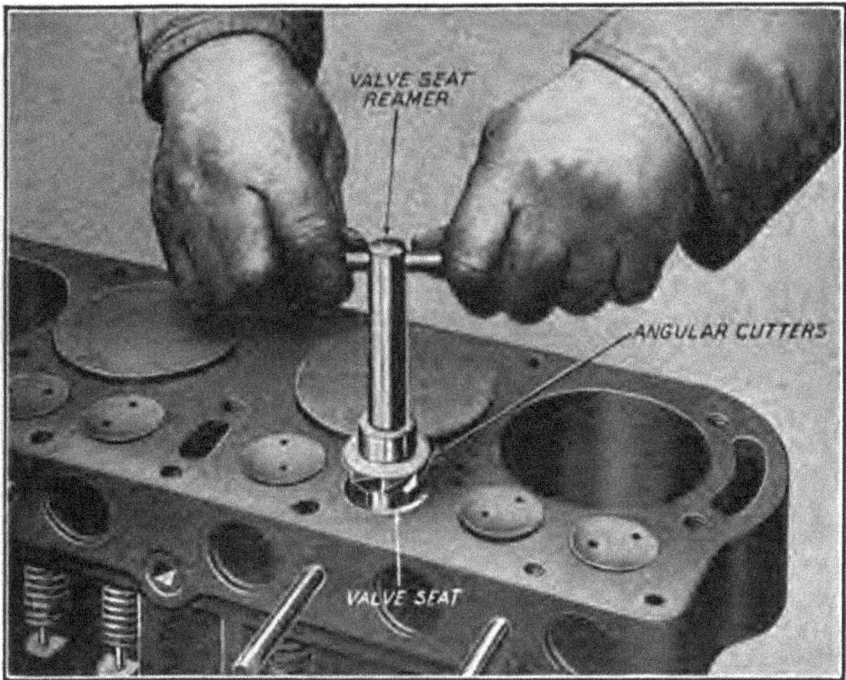

Fig. 77.—Showing Use of Valve Seat Reamer for Smoothing Scored or Pitted Valve Seats.

A special bit stock arrangement made especially for grinding in Ford valves is shown in use at Fig. 78. The lower portion of the valve grinding tool has two pins to engage the corresponding holes in the valve head. An abrasive paste is placed between the valve head and seat and the valve is turned or oscillated so the cutting material removes the roughness from the valve seat and valve, fitting one to the other. It is advisable to lift the valve from its seat frequently as the grinding operation continues, this is to provide an even distribution of the abrasive material placed between the valve head and its seat. Only sufficient pressure is given to the bit stock to overcome the uplift of the spring beneath the valve and to insure that the valve will be held against the seat. Where the spring is not used it is possible to lift the valve from time to time with the hand, which is placed under the valve stem to raise it as the grinding is carried on. It is not always possible to lift the valve in this manner when the cylinders are in place on the engine base owing to the space between the valve lift plunger and the end of the valve stem. In this event, the use of the spring as shown in Fig. 78, where it is discernible through the gas port will be desirable to raise the valve head when the grinding tool is lifted and the pressure released.

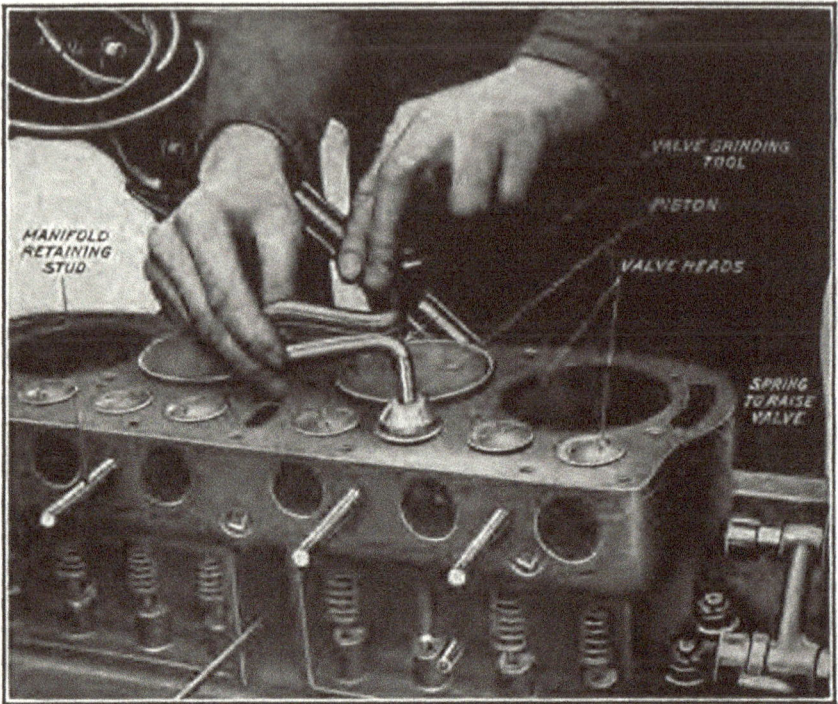

Fig. 78.—Showing Use of Special Tool for Grinding Ford Valves.

The abrasive generally used is a paste made of medium or fine emery and lard oil or kerosene. This is used until the surfaces are comparatively smooth, after which the final polish or finish is given with a paste of flour emery, grindstone dust, crocus, or ground glass and oil. An erroneous impression prevails in some quarters that the whole head surface and the seating must have a mirror-like polish. While this is not necessary it is essential that the seat in the cylinder and the bevel surface of the hand be smooth and free from pits or scratches at the completion of the operation. All traces of the emery and oil should be thoroughly washed out of the valve chamber with gasoline before the valve mechanism is assembled, and in fact it is advisable to remove the old grinding compound at regular intervals, wash the seat thoroughly and supply fresh materials from time to time as the process is in progress. The truth of seatings may be tested by taking some Prussian blue pigment and spreading a thin film of it over the valve seat. The valve is dropped in place and is given about one-eighth turn with a little pressure on the tool. If the seating is good both valve head and seat will be covered uniformly with color. If high spots exist, the heavy deposit of color will show these while the low spots will be made evident because of the lack of pigment. The grinding process should be continued until the test shows an even bearing of the valve head at all points of the cylinder seating.

Inspection of Piston Rings. — The piston rings should be taken out of the piston grooves and all carbon deposits removed from the inside of the ring and the bottom of the groove. It is important to take this deposit out because it prevents the rings from performing their proper functions by reducing the ring elasticity, and if the deposit is allowed to accumulate it may eventually result in sticking and binding of the ring, this producing excessive friction or loss of compression. When the rings are removed they should be tested to see if they retain their elasticity. If gas has been blowing by the rings or if these members have not been fitting the cylinder properly the points where the gas passed will be evidenced by a burnt, brown or roughened portion of the polished surface of the pistons and rings. The point where this discoloration will be noticed more often is at the thin end of an eccentric ring, the discoloration being present for about ½" or ¾" each side of the slot. It may be possible that the rings were not true when first put in. This made it possible for the gas to leak by in a small amount initially which increased due to continued pressure until quite a large area for gas escape had been created.

Piston Ring Manipulation. — Removing piston rings is a difficult operation if the proper means are not taken but is a simple one when the trick is known. The tools required are very simple, being three strips of thin steel about one-quarter inch wide and four or five inches long and a pair of spreading tongs made up of one-quarter inch diameter keystock tied in the center with a copper wire to form a hinge. The construction is such that when the hand is closed and the handles brought together the other end of the expander spreads out, an action just opposite to that of the conventional pliers. The method of using the tongs and the metal strips is clearly indicated at Fig. 79. At A the ring expander is shown spreading the ends of the rings sufficiently to insert the pieces of sheet metal between one of the rings and the piston. Grasp the ring as shown at B, pressing it off with the thumbs on the top of the piston and the ring will slide off easily, the thin metal strips acting as guide members to prevent the ring from catching in the other piston grooves. Usually no difficulty is experienced in removing the top or bottom rings as these members may be easily expanded and worked off directly without the use of a metal strip. When removing the intermediate rings, however, the metal strips will be found very useful. These are usually made by the repairman by grinding the teeth from old hacksaw blades and rounding the edges and comers in order to reduce the liability of cutting the fingers. By the use of the three metal strips a ring is removed without breaking or distorting it and practically no time is consumed in the operation.

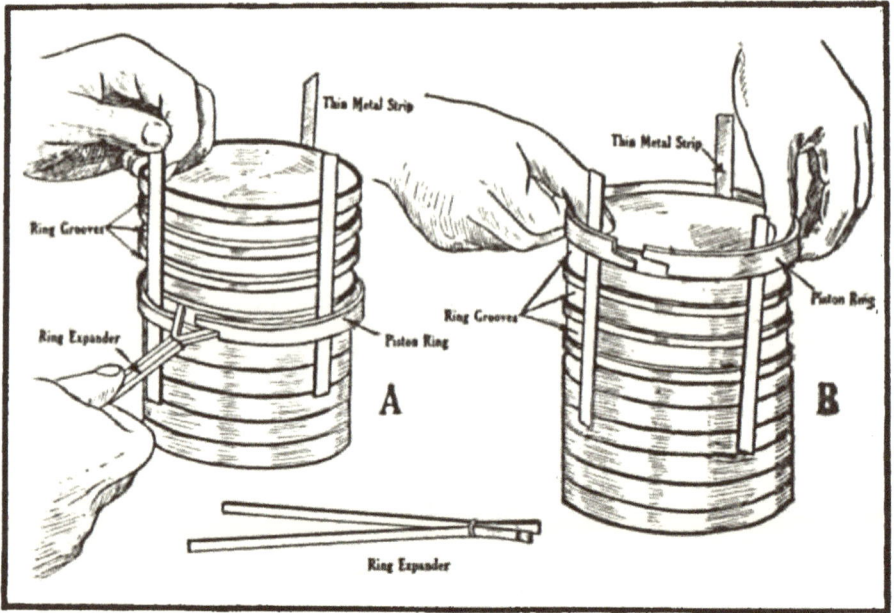

Fig. 79.—Showing Method of Removing Piston Rings Without Damaging Them.

Fitting Piston Rings. — Before installing new rings, they should be carefully fitted to the grooves to which they are applied. The tools required are a large piece of fine emery cloth, a thin flat file, a small vice with copper or leaden jaw clips and a hard smooth surface such as that afforded by the top of a surface plate or a well-planed piece of hard wood. After making sure that all deposits of burnt oil and carbon have been removed from the piston grooves, three rings are selected, one for each groove. The ring is turned all around its circumference into the groove it is to fit which can be done without springing it over the piston as the outside edge of the ring may be used to test the width of the groove just as well as the inside edge. The ring should be a fair fit and while free to move circumferentially there should be no appreciable up and down motion. If rings are a tight fit, each should be laid edge down upon the piece of emery cloth which is placed on the surface plate and carefully rubbed down until it fits the groove it is to occupy. It is advisable to fit each piston ring individually and to mark them in some way to insure that they will be placed in the groove to which they are fitted.

The repairman next turns his attention to fitting the ring in the cylinder itself. The ring should be pushed into the cylinder at least two inches up from the bottom and endeavor should be made to have the lower edge of the ring parallel with the bottom of the cylinder. If the ring is not of correct diameter

138

but is slightly larger than the cylinder bore, this condition will be evident by the angular slots of the rings being out of line or by difficulty in inserting the ring if it is a lap joint form. If such is the case the ring is removed from the cylinder and placed in the vice between soft metal jaw clips. Sufficient metal is removed with a fine file from the edges of the ring at the slot until the edges come into line and a slight space exists between them when the ring is placed in the cylinder. It is important that this space be left between the ends for if this is not done when the ring becomes heated the expansion of metal may cause the ends to abut and the ring to jam in the cylinder.

It is necessary to use more than ordinary caution in replacing the rings on the piston because they are usually made of cast iron, a metal that is very fragile and liable to break because of its brittleness. Special care should be taken in replacing new rings as these members -are more apt to break than old ones. This is probably accounted for by heating action on used rings which tends to anneal the metal as well as making it less springy. The bottom ring should be placed in position first, which is easily accomplished by springing the ring open enough to pass on the piston and then sliding it into place in the lower groove which on some types of engines is below the wrist pin whereas in others all grooves are above that member. The other members are put in by a reversal of the process outlined at Fig. 79. It is not always necessary to use the guiding strips of metal when replacing rings as it is often possible, by putting the rings on the piston a little askew and manoeuvring them to pass the grooves without springing the ring into them. The top ring should be the last one placed in position.

Before replacing pistons in the cylinder one should make sure that the slots in the piston rings are spaced equidistant on the piston. The slots should never be in line, especially with diagonal cut rings. The cylinder should be well oiled before attempt is made to install the pistons. The engine should be run with more than the ordinary amount of lubricant for several days after new piston rings have been inserted. On first starting the engine, one may be disappointed in that the compression is even less than that obtained with the old rings. This condition will soon become remedied as the rings become polished and adapt themselves to the contour of the cylinder. It will take fully 100 miles of road work to bring the rings to a sufficiently good fit so that a marked improvement in compression will be noticed.

Wrist Pin Wear. — While wrist pins are usually made of very tough steel, case-hardened, with the object of wearing out an easily renewable bronze bushing in the piston bosses rather than the wrist pin, it sometimes happens that these members will be worn so that even the replacement of new bushings in the piston will not reduce the lost motion and attendant noise due to a loose wrist pin. The only remedy is to fit new wrist pins to the piston. Where the connecting rod is clamped to the wrist pin and that member oscillates in the piston bosses the wear will usually be indicated on bronze bushings which are pressed into the piston bosses. These are easily renewed and after running a reamer of the proper size through them no difficulty should be experienced in replacing either the old or a new wrist pin depending upon the condition of that member.

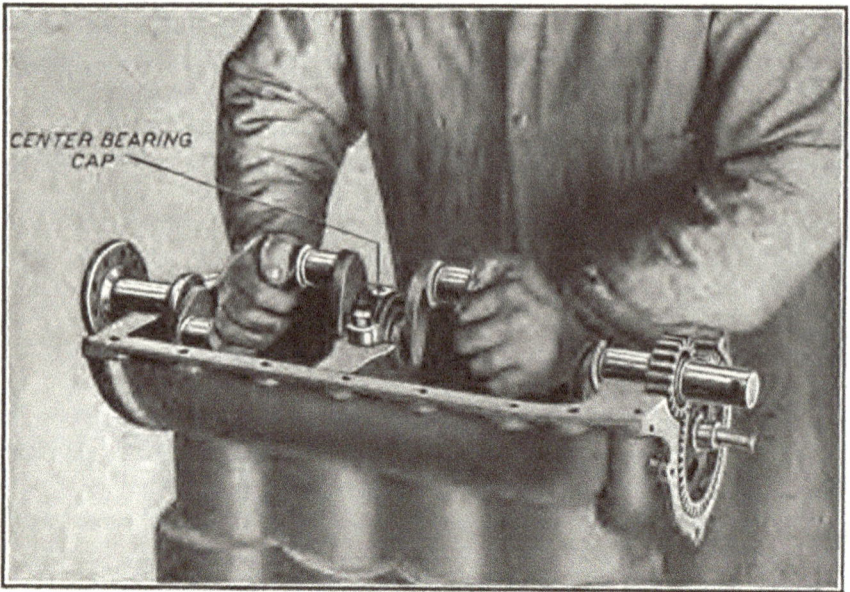

Fig. 80.—Showing Method of Testing Main Bearings When Refitting by Rocking the Crank Shaft by Hand.

Inspection and Refitting of Engine Bearings. — While the engine is dismantled one has an excellent opportunity to examine the various bearing points in the engine crank case to ascertain if any looseness exists due to depreciation of the bearing surfaces. As will be evident from Fig. 73 the three main crank shaft bearings and the lower ends of the connecting rods may be easily examined for deterioration when the crank case lower half is removed. With the rods in place as shown it is not difficult to feel the amount of lost motion by grasping the connecting rod firmly with the hand and attempting to move it up and down. The appearance of the engine base after the connecting rods and flywheel have been removed from the crank shaft is shown at Fig. 80, while the appearance of the inverted upper portion of the crank case after the crank shaft is removed is clearly shown at Fig. 81. In this view the cylinder block is supported head end down on the bench.

After the connecting rods have been removed and the flywheel taken off the crank shaft to permit of ready handling, any looseness in the main bearing may be detected by lifting up either the front or rear end of the crank shaft with a pinch bar and observing if there is any lost motion between the shaft journal and the main bearing caps. It is not necessary to take an engine entirely apart to examine the main bearings, as in the Ford engine these may be readily reached by removing a large inspection plate from the bottom of the engine crank case. This may be done without taking the engine out of the

140

frame, though if bearings are worn to any extent much time will be saved by having the engine base on a bench where it can be reached easily, and worked on with some degree of comfort.

"Knocking" Indicates Loose Bearings. — If an engine knocks when a vehicle is traveling over level roads regardless of speed or spark lever position and the trouble is not due to carbon deposits in the combustion chamber one may reasonably surmise that the main bearings have become loose or that lost motion may exist at the connecting rod big ends and possibly at the wrist pins. The main journals of the Ford engine are proportioned with ample surface and will not wear unduly unless lubrication has been neglected. The connecting rod bearings wear quicker than the main bearings owing to being subjected to a greater unit stress and it may be necessary to take these up several times in a season if the car is driven to any extent. Main bearings should run for ten thousand miles without attention in a properly built engine that has always been well oiled. Most connecting rod bearings will loosen up enough to be taken up in five thousand miles.

Adjusting Main Bearings. — When the bearings are not worn enough to require refitting the lost motion can often be eliminated by removing one or more of the thin shims or liners ordinarily used to separate the bearing caps from the seat. Care must be taken that an even number of shims of the same thickness are removed from each side of the journal. If there is considerable lost motion after one or two shims have been removed it will be advisable to file some metal from the bearing cap and to scrape the bearing to a fit before the bearing cap is finally tightened up.

The following instructions for refitting main bearings are given by the Ford Motor Company in their book of instructions:

(1) After the engine has been taken out of the car, remove crank case, transmission cover, cylinder head, pistons, connecting rods, transmission and magneto coils. Take off the three babbitted caps and clean the bearing surfaces with gasoline. Apply blue or red lead to the crank shaft bearing surfaces, which will enable you, in fitting the caps, to determine whether a perfect bearing surface is obtained.

(2) Place the rear cap in position and tighten it up as much as possible without stripping the bolt threads. When the bearing has been properly fitted, the crank shaft will permit moving with one hand. If the crank shaft cannot be turned with one hand, the contact between the bearing surfaces is evidently too close, and the cap requires shimming up, one or two brass liners usually being sufficient. In case the crank shaft moves too easily with one hand, the shims should be removed and the steel surface of the cap filed off, permitting it to set closer.

(3) After removing the cap, observe whether the blue or red "spottings" indicate a full bearing the length of the cap. If "spottings" do not show a true bearing, the babbitt should be scraped and the cap refitted until the proper results are obtained.

(4) Lay the rear cap aside and proceed to adjust the center bearing in the same manner. Repeat the operation with the front bearing, with the other two bearings laid aside.

(5) When the proper adjustment of each bearing has been obtained, clean the babbitt surface carefully and place a little lubricating oil on the bearings, also on the crank shaft; then draw the caps up as closely as possible — the necessary shims, of course, being in place. Do not be afraid of getting the cap bolts too tight, as the shim under the cap and the oil between the bearing surfaces will prevent the metal being drawn into too close contact. If oil is not put on the bearing surfaces, the babbitt is apt to cut out when the motor is started up before the oil in the crank case can get into the bearing.

In replacing the crank case and transmission cover on the motor, it is advisable to use a new set of felt gaskets to prevent oil leaks.

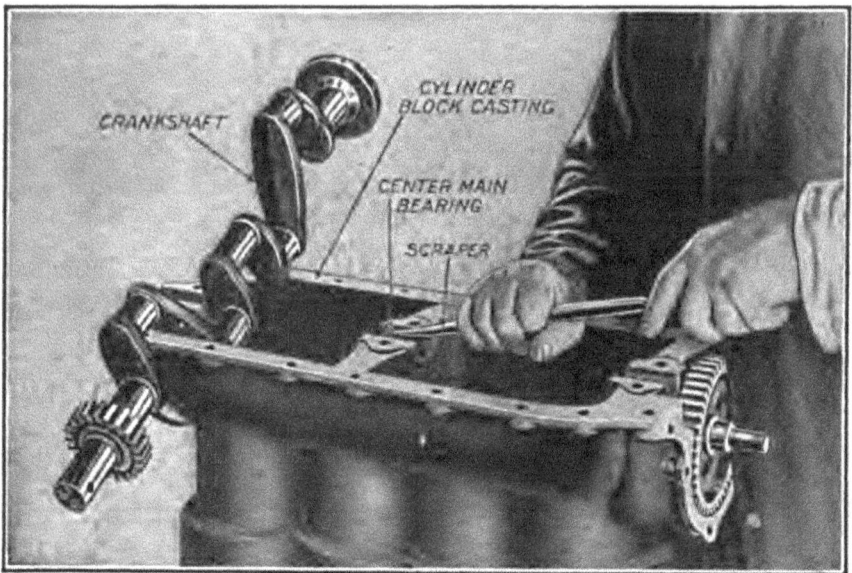

Fig. 81.—Showing Method of Scraping in Main Bearings to Fit Crank Shaft Journals.

Scraping Brasses to Fit. — To insure that the bearing brasses will be a good fit on the crank pins or crank shaft journals they must be scraped to fit. The process of scraping, while a tedious one, is not difficult, requiring only patience and some degree of care to do a good job. The crank pin surface is smeared with Prussian blue pigment which is spread evenly over the entire surface. The bearings are then clamped together in the usual manner with proper bolts and the crank shaft revolved several times to indicate the high spots on the bearing cap. At the start of the process of scraping in, the bearing may seat at only a few points. Continued scraping will bring the bearing surface practically across the brass, which is a considerable improvement,

142

while the process may be considered complete when the brass indicates a bearing all over. The high spots are indicated by bine, as where the shaft does not bear on the bearing there is ne color. The high spots are removed by means of scraping tools of the form shown at Figs. 81 and 85, which may be easily made from womont files. These are forged to shape and ground hollow and are kept properly sharpened by frequent rubbing on an ordinary oil stone. To scrape properly the edge of the scraper must be very keen. The straight or curved half round type works well on soft bearing metals such as babbitt or white brass, but on yellow brass or bronze it cuts very slowly and as soon as the edge becomes dull considerable pressure is needed to remove any metal, this calling for frequent sharpening.

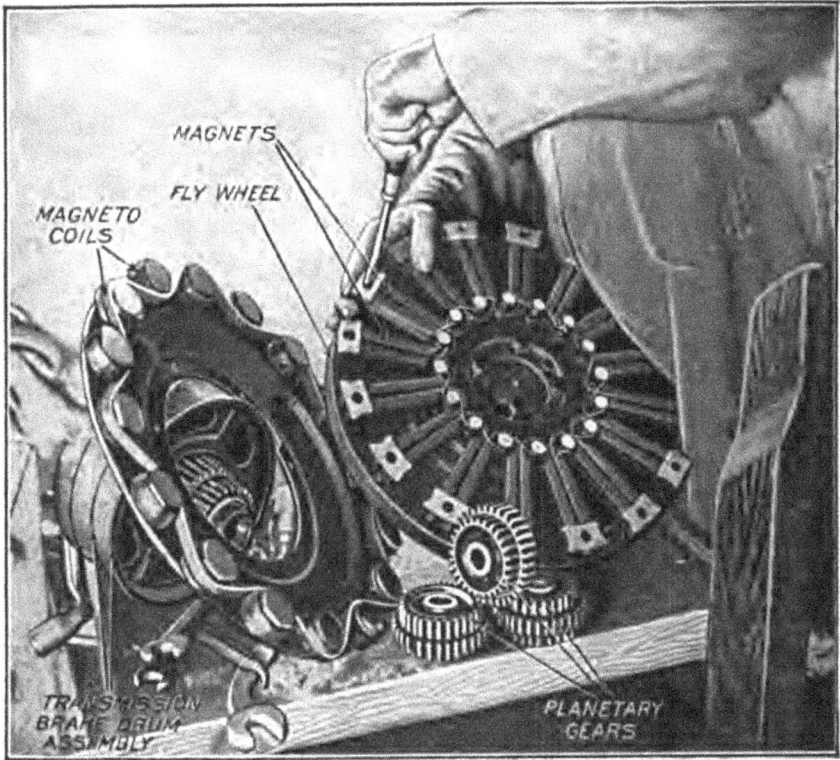

Fig. 82.—Method of Fastening Magnets in Place on Flywheel. Note Planetary Triple Gear Assemblies of Transmission in Foreground.

When correcting errors on flat or curved surfaces by handscraping, it is desirable, of course, to obtain an evenly spotted bearing with as little scraping as possible. When the part to be scraped is first applied to the surface-plate, or to a journal in the case of a bearing, three or four "high" spots may be indicated by the marking material. The time required to reduce these high spots and obtain a bearing that is distributed over the entire surface depends largely upon the way the scraping is started. If the first hearing marks indicate a decided rise in the surface, much time can be saved by scraping larger areas than covered by the bearing marks, this is especially true of engine bearings. An experienced workman will not only remove the heavy marks, but also reduce a larger area; then, when the bearing is tested again, the marks will generally be distributed somewhat. If the heavy marks which usually appear at first are simply removed by light scraping, these "point bearings" are gradually enlarged, but a much longer time will be required to distribute them.

The number of times the bearing must be applied to the journal for testing is important. The time required to distribute the bearing marks evenly depends largely upon one's judgment in "reading" these marks. In the early stages of the scraping operation the marks should be used partly as a guide for showing the high areas, and instead of merely scraping the marked spot the surface surrounding it should also be reduced, unless it is evident that the unevenness is local. The idea should be to obtain first a few large but generally distributed marks; then an evenly and finely spotted surface can be produced quite easily.

Remetalling and Fitting Connecting Bods. — Fitting and adjusting rod bearings, especially those at the crank pin end is one of the operations that must be performed several times a season if a car is used to any extent. There are two forms of connecting rods in general use, known respectively as the marine type, and the hinged form. The hinge type is the simplest, but one clamp bolt being used to keep the parts together as the cap is hinged to the rod end on one side, this permitting the lower portion to swing down and the crank pin to pass out from between the halves when the retaining bolt is removed. In the marine type, which is that used on the Ford Model T, one bolt is employed at each side and the cap must be removed entirely before the bearing can be taken off of the crank pin. The tightness of the brasses around the crank pin can never be determined solely by the adjustment of the bolts, as while it is important that these should be drawn up as tightly as possible the bearing should fit the shaft without undue binding even if the brasses must be scraped to insure a proper fit. As is true of the main bearings the marine form of connecting rod may have a number of liners or shims interposed between the top and cap portions of the rod end, and these may be reduced in number when necessary to bring the brasses closer together.

Before assembling on the shaft, it is necessary to fit the bearings by scraping, the same instructions given for restoring the contour of the main bearings applying just as well in this case. It is apparent that if the crank pins are not round no amount of scraping will insure a true bearing. A point to observe is to make sure that the heads of the cap retaining bolts are imbedded solidly in their proper position and that they are not raised by any burrs or particles of dirt under the head which will flatten out after the engine has been run for a time and allow the bolts to slack off. Similarly, care should be taken that there is no foreign matter under the brasses and the box in which they seat. To guard against this, the bolts should be struck with a hammer several times after they are tightened up and the connecting rod can be hit sharply several times under the cap with a wooden mallet or lead hammer.

Care should be taken in screwing on the retaining nuts to insure that they will remain in place and not slack off. Spring washers should never be used on either connecting rod ends or main bearing bolts because these sometimes snap in two pieces and leave the nut slack. The best method of locking is to use well-fitting split pins and castellated nuts as supplied.

The following advice from the Ford Manual is pertinent:

"Remember, there is a possibility of getting the bearings too tight, and under such conditions the babbitt is apt to cut out quickly, unless precaution is taken to run the motor slowly at the start. It is a good plan after adjusting the bearings to jack up the rear wheels and let the motor run slowly for about two hours (keeping it well supplied with water and oil) before taking it out on the road. Whenever possible these bearings should be fitted by an expert Ford mechanic."

If the babbitt lining is worn sufficiently so it must be replaced or if it has been burned out by running the engine without oil, the connecting rod and cap must be rebabbitted. The Ford makers advise returning the connecting rods to the factory or to the nearest service station. They advise as follows:

"Worn connecting rods may be returned, prepaid, to the nearest agent or branch house for exchange at a price of 75 cents each to cover the cost of rebabbitting. It is not advisable for any owner or repair shop to attempt the rebabbitting of connecting rods or main bearings, for without a special jig in which to form the bearings, satisfactory results will not be obtained. The constant tapping of a loose connecting rod on the crank shaft will eventually produce crystallization of the steel — result, broken crank shaft and possibly other parts of the engine damaged."

If the parts are wanted in a hurry, it is possible for any competent repair man to rig up and replace the lining metal himself, though this will not be a profitable operation if the parts can be procured from the factory in time.

The repair man who is called upon to replace the bearing metal will find the following instructions regarding remetalling bearings of value. The method described was used by the writer while in charge of a large shop where much work of this kind was done, and while instructions given apply specifically to lining the big ends of connecting rods the same process may be used successfully on any other bearings where the mandrel and collars can be used, the dimensions being changed to suit the requirements of the worker. Obviously the old metal must be thoroughly cleaned out and rod end made ready to receive the new lining before any attempt is made to pour in new metal.

In the case mentioned the journals of the crank shaft were two inches in diameter and the big ends of the connecting rods were worn too much to allow of adjusting. A piece of pipe about 9 inches long was procured and turned down in a lathe until it was a shade under 2" in diameter, which made a hollow mandrel of it. A piece of steel tubing could have been used to as good advantage had any been available. As the outside of the bearing caps were machined true a couple of set collars were bored out to be a good fit on the mandrel, and while still in the lathe they were recessed out to just fit over the outside of the big ends, as shown in sketch. Fig. 83. One of these collars was placed on the hollow mandrel. A, after which the mandrel was pushed through the big end, and the other collar was put on the other side, insuring that the mandrel was as near the center as possible for it to be. The assemblage is then supported on a couple of V-blocks which are supported on a lathe bed, the ends of the mandrel lying within the V's, while the connecting rod hangs between the ways. A piece of solid round iron or steel which will go inside of the hollow mandrel should be made red hot while the anti-friction metal is being melted and is pushed inside the mandrel to heat it. In a minute or two the metal may be poured in through B to fill annulus D, and as the metal and the big end caps are well heated the molten metal will flow to every point.

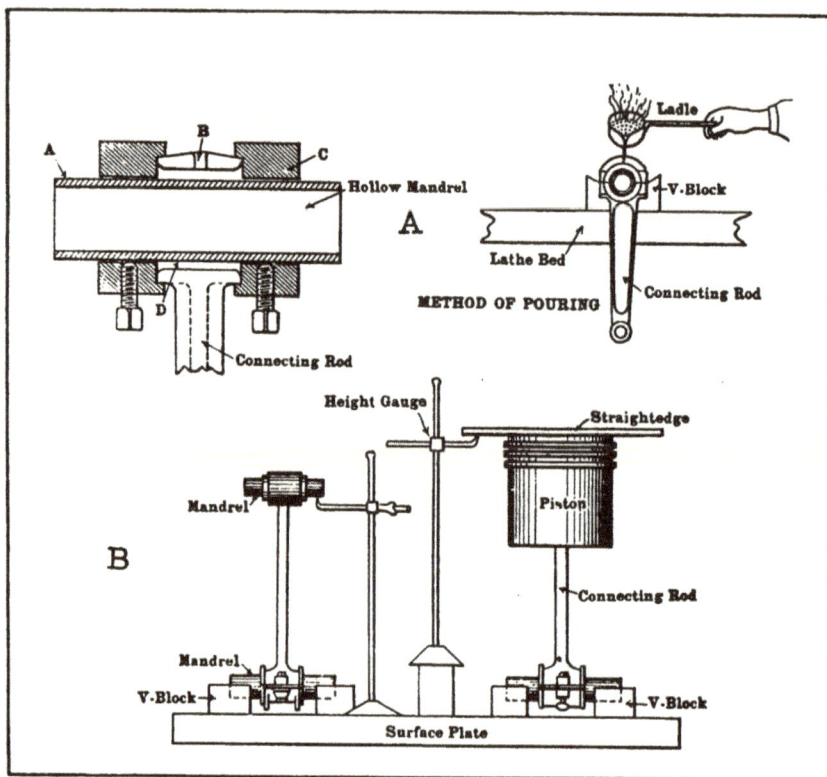

Fig. 83.—Illustrations Showing Method of Rebabbitting Connecting Rod Bearings at A and Method of Testing Connecting Rod Bearing Parallelism at B.

The heating of the mandrel can be just as well accomplished by directing the flame from a blow torch or Bunsen burner into the opening. After the metal is poured and has set well the whole may be easily cooled by directing a blast of air against the big end. During the pouring process the cap is separated from the rod end by shims of oiled cardboard. This is afterward replaced by brass shims. As is evident, the thinner the liners and the greater the number used, the more sensitive the character of future adjustment possible. A hollow mandrel is to be preferred to a solid one because of the ease with which it can be heated and cooled. The mandrel should be about .025" smaller than the crank pins. Vents should be made for the heated gases by grooving the face of each of the collars nearest the big end and on the same side as the hole through which the metal is poured. If provision is not made for "venting" the molten metal will not run uniformly and will become honeycombed. After cooling, the bearing is either bored out in a lathe to the size of the journal or scraped to a fit by hand. The method of pouring the molten metal is clearly shown while the sectional view makes the construction and application of the mandrel clear. The same method may be used to rebabbitt main boxes except that a pair of collars will be needed for each bearing and a long mandrel used to insure proper alignment of the three bearings.

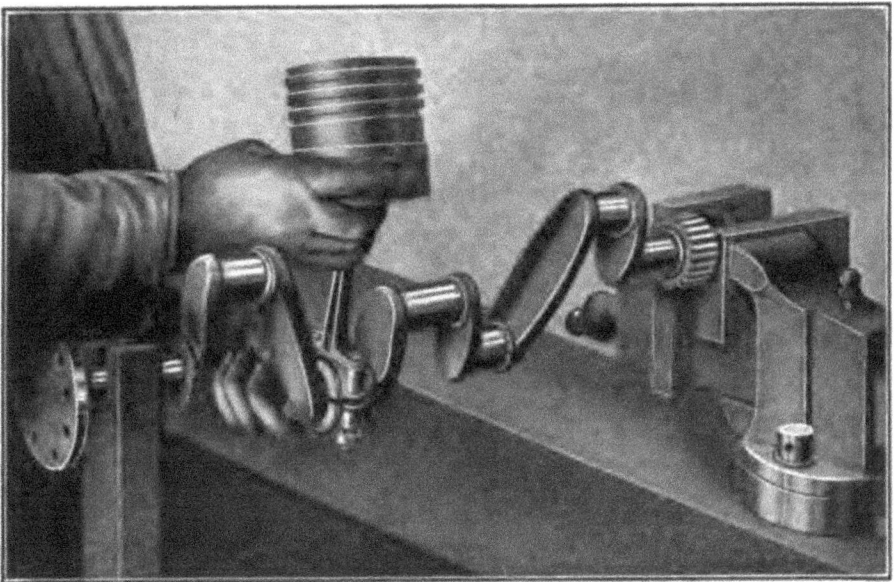

Fig. 84.—Connecting Rod Bearings May be Easily Fitted to the Crank Shaft if This Member is Removed from the Engine and Supported by Bench Vise.

Testing Bearing Parallelism. — It is not possible to give other than general directions regarding the proper degree of tightening for a connecting rod bearing, but as a guide to correct adjustment it may be said that if the connecting rod cap is tightened sufficiently so the connecting rod will just about fall over from a vertical position due to the piston weight when the

147

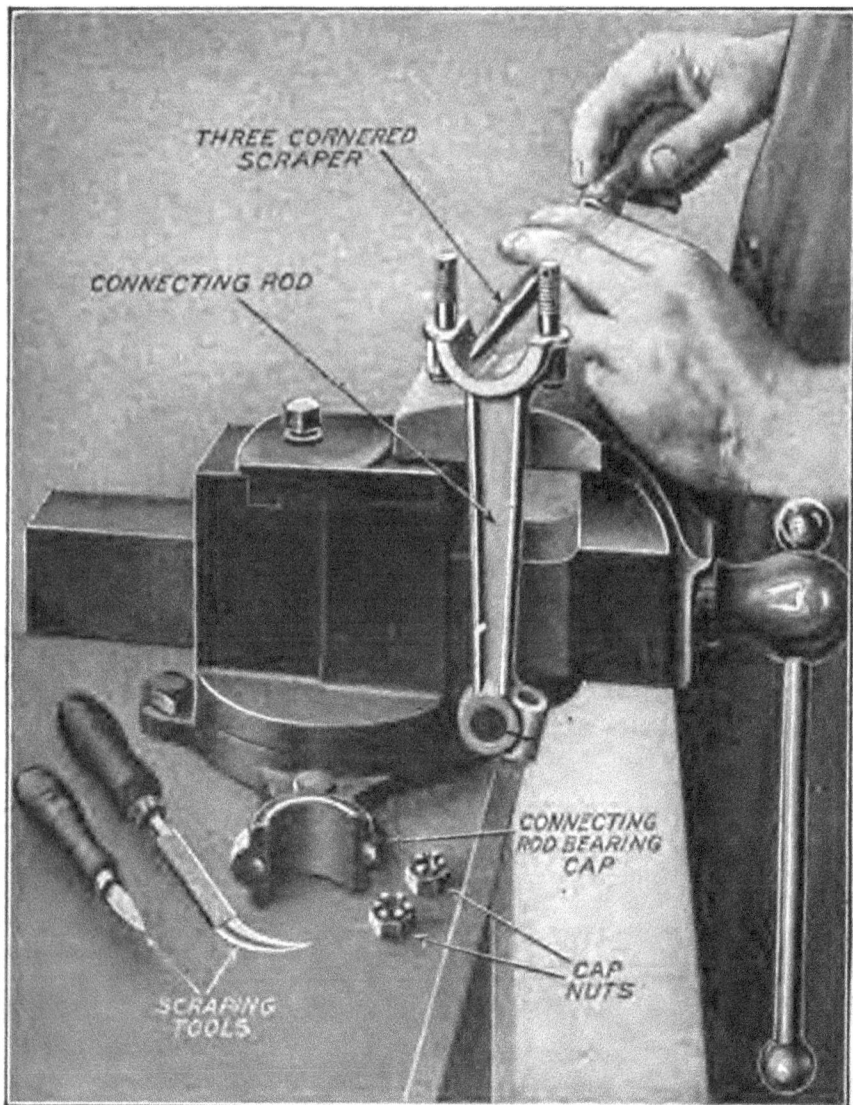

Fig. 85.—Illustrating Method of Scraping In Connecting Rod Bearings and Tools Used in This Process.

bolts are fully tightened up as in Fig. 84, the adjustment will be nearly correct. As previously stated, babbitt or white metal bearings can be set up more tightly than bronze, as the metal is softer and any high spots will soon be leveled down with the running of the engine. It is important that care be taken to preserve parallelism of the wrist pins and crank shaft while scraping in bearings. This can be determined in several ways. That shown at Fig. 83, B, is used when the parts are not in the engine assembly and when the connecting rod bearing is being fitted to a mandrel or arbor the same size as the crank pin. The arbor, which is finished very smooth and of uniform diameter, is placed in two V-blocks which in turn are supported by a level surface plate. An adjustable height gauge may be tried, first at one side of the wrist pin which is placed at the upper end of the connecting rod, then at the other and any variation will be easily determined by the degree of tilting of the rod. This test may be made with the wrist pin alone or if the piston is in place a straight edge on the piston top may be employed. A spirit level will readily show any inclination, while the straight edge is used in connection with the height gauge as indicated.

Camshafts and Timing Gears. — Knocking sounds are also evident if the camshaft is loose in its bearings and also if the timing gears are loose on the shaft. The camshaft is usually supported by solid bearings of the removable bushing type having no compensation for depreciation. If these bearings wear the only remedy is replacement with new ones. Another point to watch is the method of retaining the camshaft gear in place. On some engines the gear is fastened to a flange on the camshaft by retaining screw. These are not apt to become loose, but where reliance is placed on a key, as in the Ford, the camshaft gear may often be loose on its supporting member. The only remedy is to enlarge the key slot in both gear and shaft and to fit a larger retaining key.

If the camshaft is sprung or twisted it will alter the valve timing to such an extent that the smoothness of operation of the engine will be materially affected. If this condition is suspected the cam shaft may be swung on lathe centers and turned to see if it runs out and if bent it can be straightened in any of the usual forms of shaft straightening machines. The shaft may be twisted without being sprung. This can only be determined by supporting one end of the shaft in an index head and the other end on a milling machine center. The cams are then checked to see that they are separated by the proper degree of angularity. This process is one that requires a thorough knowledge of the valve timing of the engine in question and is best done at the factory where the engine was made. The timing gears should also be examined to see if the teeth are worn enough so that considerable back lash or lost motion exists between them. A worn timing gear not only produces noise but it will cause the time of opening and closing of the engine valves to vary materially.

Valve Timing Method. — Among the factors making for efficient operation of the gasoline engine, especially of the multiple cylinder type used for automobile propulsion, there is none of more importance than proper valve timing. In the Ford four cylinder four-cycle motor there are eight of these members, two to each cylinder, the function of the inlet valves being to permit the cylinders to fill with gas while the exhaust valves open to clear the cylinders of the products of combustion. The inlet valve usually opens when the piston is at approximately the top of its stroke in the cylinder or during that portion of the engine cycle where the piston is starting to go down to draw in a charge of gas. This valve is opened at a period equal to the downstroke of the piston and sometimes more, but is closed during the succeeding compression, explosion and scavenging strokes. The operation of the exhaust valve is very much the same as the inlet except that it is opened for a longer period, starting to open before the piston has completed the downward stroke produced by the explosion and is sometimes opened slightly after the end of the return or scavenging stroke.

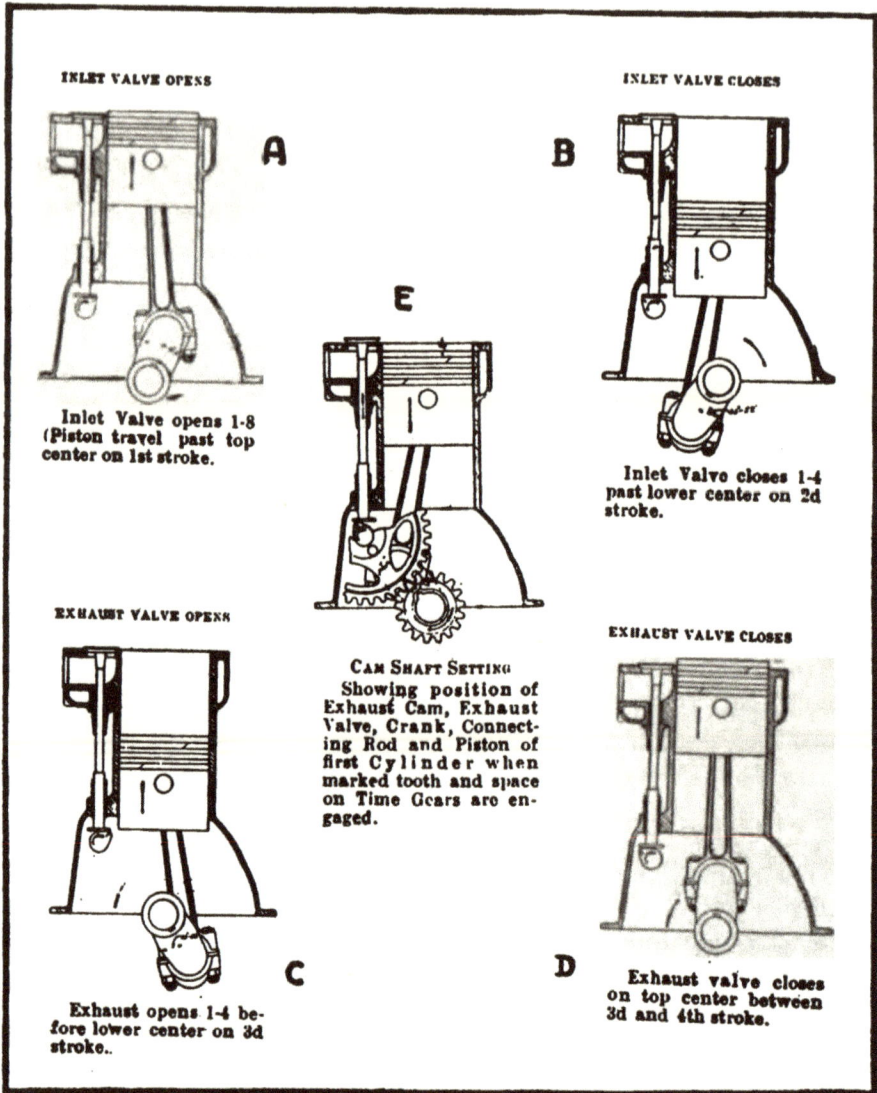

INLET VALVE OPENS

A

Inlet Valve opens 1-8 (Piston travel past top center on 1st stroke.

INLET VALVE CLOSES

B

Inlet Valve closes 1-4 past lower center on 2d stroke.

E

CAM SHAFT SETTING
Showing position of Exhaust Cam, Exhaust Valve, Crank, Connecting Rod and Piston of first Cylinder when marked tooth and space on Time Gears are engaged.

EXHAUST VALVE OPENS

C

Exhaust opens 1-4 before lower center on 3d stroke..

EXHAUST VALVE CLOSES

D

Exhaust valve closes on top center between 3d and 4th stroke.

Fig. 86.—Diagram Showing Method of Timing Ford Valves.

When the space between the valve stem and the valve lifter is more than it should be there are two methods of compensating for this depreciation. On many small motors no adjustment is provided between the valve stem and the valve stem plunger. The makers of the Ford car formerly advised drawing the valve stem out until the proper space existed between the push rod and the stem. It is important when drawing out the stem or lengthening it not to bend the valve stem as this will result in the valves sticking, or in any event the bore of the valve stem guide in the cylinder will be worn unevenly. The clearance between the pushrod and the valve stem should never be greater than 1/32" nor less than 1/64". If too much clearance is present the valve will open late and close early. If the clearance is less than the minimum there is danger of the valve remaining partially open all of the time because the valve stem lengthens due to expansion produced by the heat of the explosion. When it is necessary to draw down a valve stem this should be done by peening it for about ¾" above the pinhole or key slot.

It is not a difficult matter to set the clearance exactly as it should be on those types of engines provided with an adjustment screw which may be raised or lowered in the valve plunger or, in forms having fiber inserts in the top of the valve plunger. These inserts are utilized to silence the valve action and may be easily removed and replaced with new ones when worn. A simple and cheap accessory that can be obtained on the open market can be used to adjust the clearance on Ford and similar type motors. This consists of a number of stamped steel cups that can be pushed on the lower portion of the valve stem and a number of thin steel washers to be interposed inside of the cup and between the bottom of that member and the end of the valve stem to regulate the clearance as desired.

Valve Timing in Ford Engines. — After the valves have been ground in and seated properly it is important that the timing of the valves be verified. As the valves have been properly timed when the engine is assembled at the factory the only reason for checking the timing would be during the overhauling when the camshaft or timing gears have been removed from the engine base. In fitting the large timing gear to the camshaft it is important to see that the first cam points in a direction opposite from the zero mark as shown at Fig. 86, E. The large and small time gears must also mesh, so that the tooth marked on the small timing gear will coincide with a similar mark between the two teeth on the large gear. With the timing gears set as indicated the exhaust valve in No. 1 cylinder is opened and the intake valve closed.

The opening and closing point of the valves is as follows: The intake valve opens with the piston Vie of an inch down from top center as shown at Fig. 86, A. The inlet valve closes 9/16 inch after the piston has reached bottom center as shown at Fig. 86, B, the distance from the top of the piston to the top of the cylinder casting measuring 31/8th inches. The exhaust valve opens when the piston reaches a point on its travel from 5/16th inch to ¼ inch before lower center on third stroke, as shown at Fig. 86, C. The distance from the top of the piston head to the top of the cylinder casting at the time the exhaust valve starts to open is 3 3/8 inches. The exhaust valve should close on top center between the third and fourth strokes as shown at Fig. 86, B. The piston top at this time is %6 inch above the cylinder casting. The clearance between the push rod and valve stem should be carefully gauged as previously mentioned. Obviously this gap should be measured when the push rod is at the extreme lower point of its travel or riding on the rounded portion and not the point of the cam.

When the push rods or valve stems become worn so as to leave too much play between them, it is best to replace with new push rods. The operation of drawing down the valve stems requires considerable experience and the price of the new part does not warrant the time and expense necessary to do the work as it should be done. Mention has been made of simple clearance adjusters made especially for Ford valves but these are not marketed by the Ford people and can only be obtained from supply houses. If the valves fail to seat themselves properly there is a possibility that the valve springs may be weak or broken. It is stated that a weak inlet valve spring will not affect the running of the engine as much as a sluggish acting exhaust valve spring. Weakness in a valve spring can be easily detected by removing the cover plate and inserting a screw driver between the coils of the spring while the engine is running. Each of the exhaust valve springs is tried in turn, and if the extra tension thus produced results in the engine picking up speed the spring is too weak and should be replaced with a new one.

Repairing Ford Magneto. — While the engine is dismantled it is a good time to make any repairs that may be necessary to the magneto, these being rarely needed on account of the simplicity of the device. The Ford magneto consists of permanent magnets, and there is not much liability of their losing their strength unless acted upon by some outside force, though cases have been known where much of the magnetism has been lost due to overheating the engine. If a dry or storage battery has been attached to the magneto terminal by mistake the magnet strength may be weakened. If on testing with a special instrument made for the purpose, the magneto does not show the proper current, instead of attempting to recharge the magnets the best and cheapest way is to install a complete set of new ones. These may be obtained from the factory or nearest agent and will be placed on a board in exactly the same manner as they should be installed on the flywheel. The magnets may be easily removed from the flywheel as outlined at Fig. 82. Great care should be taken when assembling the magnets and lining up the parts so that the pole pieces of the magnets will be separated from the surface of the coil spools carried by the stationary plate by no more than H2nd of an inch. To take the old magnets from the flywheel simply remove the cap screw and bronze screw which holds each in place.

The magneto is often blamed when the trouble is a weak current that results by waste or other foreign matter accumulating under the current collecting contact plunger which is held in place by the binding post carrier on top of the transmission case cover. Remove the three screws which serve to retain this collecting terminal in place, this allows the binding post to spring up and plunger to be withdrawn. The contact pad on top of the fixed coil assembly should be thoroughly cleaned and the end of the plunger brightened with emery cloth before it is replaced. The magneto cannot be reached for replacing magnets without taking the power plant out of the car. After the crank case and transmission covers are off, the flywheel may be taken off of the flange on the end of the crank shaft by removing four cap screws that hold the flywheel to that member. Whenever repairs are necessary to the magneto such as replacing magnets or making sure that all magnet retaining screws and bolts are tight, the best way is to have the parts on a bench where they may be easily reached, as shown at Fig. 82.

Packings and Gaskets for Ford Motor. — If the power plant has been taken apart it may be found desirable to replace the various gaskets and packing members with new ones when reassembling the parts. This applies especially to the felt packings, which are apt to be torn when the pieces rest-

ing on them are removed quickly. The copper asbestos gaskets which are used as packings on the combustion chamber end of the engine do not depreciate as readily as those made of felt, though when these flatten out it

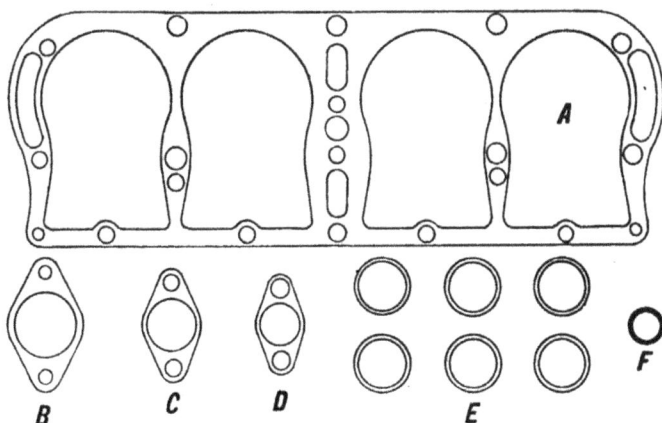

Fig. 87.—Copper Asbestos Gaskets Used on Ford Motor.

may be well to replace them with new, because they are so cheap that it would be questionable economy to use them if there was any doubt about their condition.

A set of the copper-asbestos packings for the cylinder heads of the Ford engine is shown at Fig. 87. That at A, is the member placed between the cylinder head and cylinder block, those at B and C are used between the water connections, those at E are placed under the inlet and exhaust manifolds; that at D is placed between the carburettor and flange on the lower portion of the inlet manifold, while the small packing at F is used at the lower portion of the crank case or flywheel compartment as a seating for the oil drain plug.

The felt packings are clearly shown at Fig. 88. The strips at A, are placed between the pressed steel crank case member and the cylinder block. The piece marked M, is employed between the top of the transmission case and the pressed steel lower portion. The large packing, C, is placed between the bottom plate of the crank case and the pressed steel lower crank case member. Packing D is utilized under the transmission case cover plate on late model Ford cars, while that shown at E is used on some of the earlier models having a square cover plate.

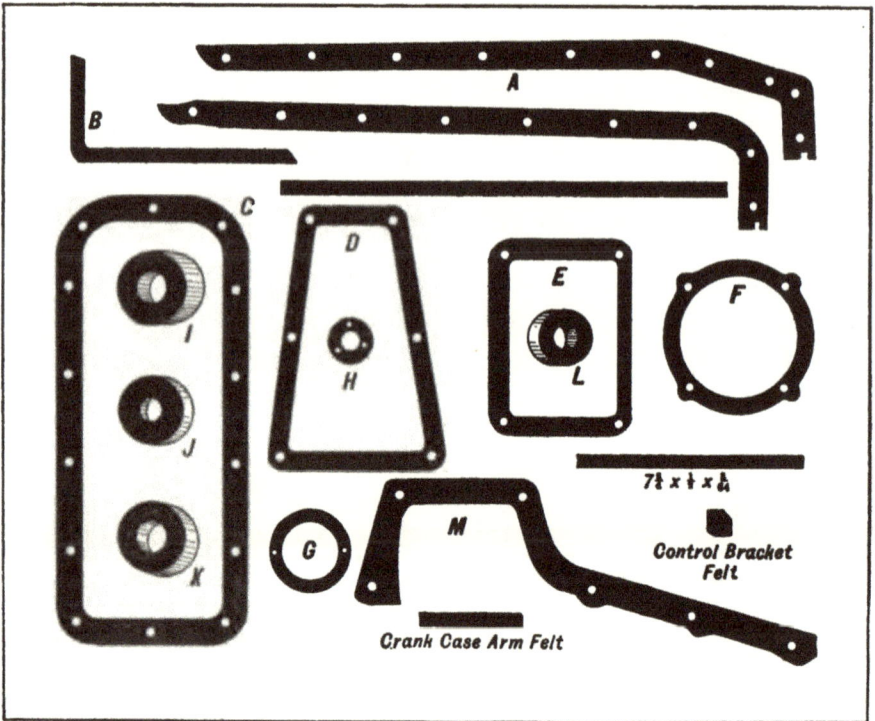

Fig. 88.—Group Showing Felt Packings for Ford Power Plant and Also for Retaining Oil in Running Gear Parts.

The small corner pieces, B, are used on the valve chamber cover plates. The gasket, F, goes between the members of the universal joint ball housing. Piece H fits under the magneto collecting brush terminal fitting. Washers I, J, K and L, are oil retention members for the front and rear axles. The control bracket felts are placed in recesses in the supporting bearing of the cross shaft to which the hand brake control lever is fastened and serve to retain oil and insure adequate lubricity of the cross shaft.

The crank case arm felts are intended to be placed between the arms and the frame side member. The strips shown which are not lettered are used to complete the packing between the timing gear case portions and between the upper transmission case and round portion against which it fits on the engine base. When replacing either the copper asbestos or felt packings it is well to coat these with heavy shellac before the parts are assembled. The shellac fills any irregularities that may exist in the surfaces and prevents oil leakage.

Precautions in Reassembling Parts. — When all of the essential components of a power plant have been carefully looked over and cleaned and all

154

defects eliminated, either by adjustment or replacement of worn portions, the motor should be reassembled, taking care to have the parts occupy just the same relative positions they did before the motor was dismantled. As each part is added to the assemblage, care should be taken to insure adequate lubrication of all new points of bearing by squirting liberal quantities of cylinder oil upon them with a hand oil can or syringe provided for the purpose. In adjusting the crank shaft bearings tighten them one at a time and revolve the shaft each time one of the bearing caps is set up to insure that the newly adjusted bearing does not have undue friction. All retaining keys and pins must be positively placed and it is good practice to cover such a part with lubricant before replacing it because it will not only drive easier but the part may be removed more easily if necessary at some future time if no rust collects around it.

When a piece is held by more than one bolt or screw, especially if it is a casting of brittle material such as cast iron, the fastening bolts should be tightened uniformly. If one bolt is tightened more than the rest it is liable to spring the casting enough to break it. This applies especially to the Ford water connection castings. Spring washers, check nuts, split pins or other locking means should always be provided, especially on parts which are in motion or subjected to a load. When reassembling the inlet and exhaust manifolds it is well to use only perfect packings or gaskets and to avoid the use of those that seem to have hardened up or flattened out too much in service. If it is necessary to use new gaskets it is imperative to employ these at all joints on manifold because if old and new gaskets are used together the new ones are apt to keep the manifold from bedding properly upon the used ones.

It is well to coat the threads of all bolts and screws subjected to heat, such as cylinder head bolts and exhaust pipe retaining nut, with a mixture of graphite and oil. Those that enter the water jacket should be covered with white or read lead or pipe thread compound. Gaskets will hold better if coated with shellac before manifold or other parts are placed over them. The shellac fills any irregularities in the joints and assists materially in preventing leakage after the joint is made up and the coating has a chance to set. In replacing cylinder head packings on cars like the Ford it is well to run the engine for a short while, several minutes at the most, without any water in the jacket in order to heat the head up thoroughly. It will usually be found possible to tighten down a little more on all of the cylinder head retaining bolts after this is done because if the gasket has been coated with shellac the surplus material will have burnt off and the entire packing bedded down. Care should be taken when using shellac, white or red lead, etc., not to supply so much that the surplus will run into the cylinder, water jacket or gas passages.

Taking Down Transmission. — After the car has been in use for a considerable period, especially in sections of the country where frequent use of the slow speed gearing is necessary, the transmission will become very noisy in action and will rattle and grind when either the slow or reverse brake

155

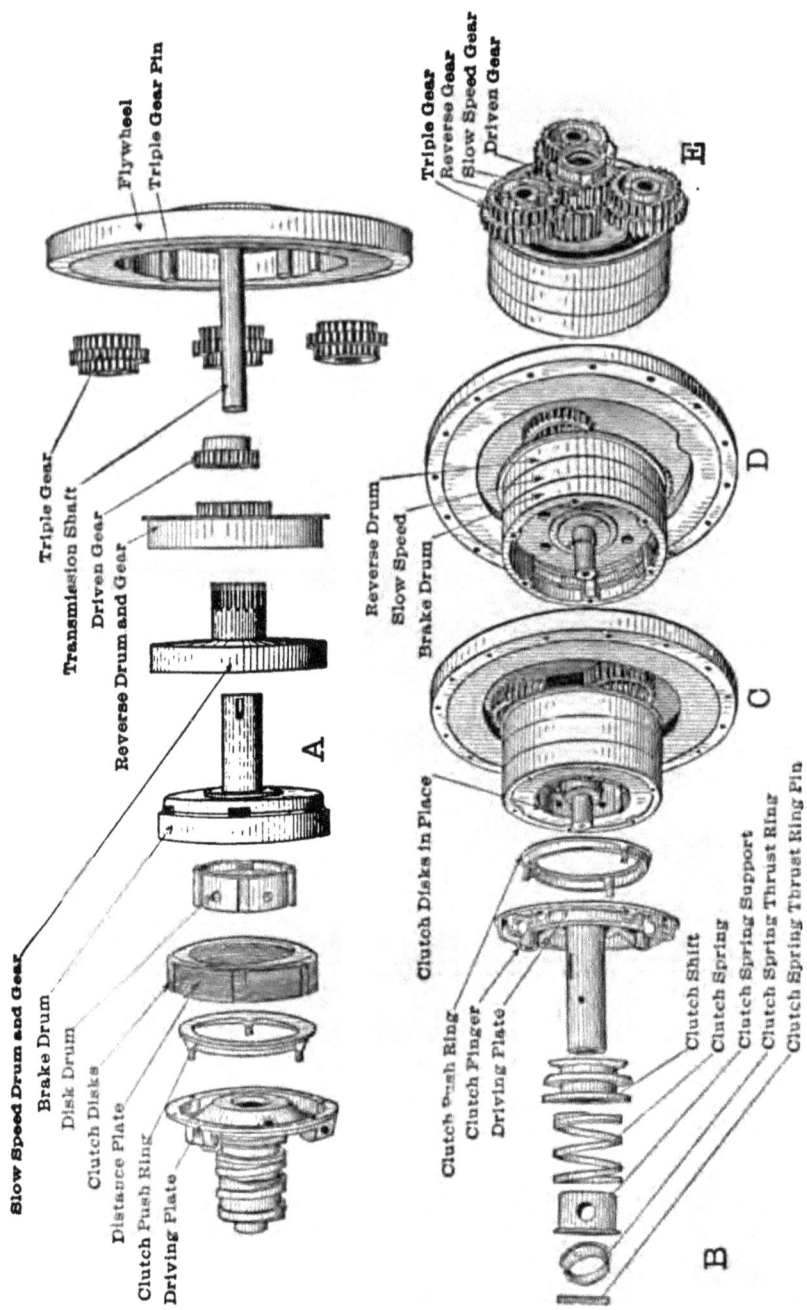

Flywheel
Triple Gear Pin

Triple Gear
Reverse Gear
Slow Speed Gear
Driven Gear

Triple Gear
Transmission Shaft
Driven Gear
Reverse Drum and Gear

Reverse Drum
Slow Speed
Brake Drum

Slow Speed Drum and Gear
Brake Drum
Disk Drum
Clutch Disks
Distance Plate
Clutch Push Ring
Driving Plate

Clutch Disks in Place

Clutch Push Ring
Clutch Finger
Driving Plate

Clutch Shift
Clutch Spring
Clutch Spring Support
Clutch Spring Thrust Ring
Clutch Spring Thrust Ring Pin

A
B
C
D
E

Fig. 89.—Plate Showing Parts Comprising the Ford Transmission When Disassembled at A, and When Joined Together to Form Various Groups to Facilitate Assembly at C, D and E.

bands are operated. Owing to the fact that there are no gears in operation on the high speed but little trouble will be experienced with the clutch disc assembly unless the disc surfaces have become roughened or have worn enough so that no further adjustment is possible with a clutch adjusting

156

screw. While the power plant is out of the frame it will be found desirable to take the transmission gearing apart and examine the various fastenings and gears. The gears are not apt to wear very much and practically the only trouble will be depreciation of the bushings that form bearings for the triple gear assembly and the various brake drum members.

In taking the transmission gear apart, the first thing is to drive out the clutch spring thrust ring pin which is shown at Fig. 89, B. This releases the clutch spring thrust ring and the clutch spring supports and makes it possible to remove the clutch spring, clutch shift collar and driving plate. The driving plate can only be taken off after the screws by which it is bolted to the brake drum and clutch carrier are taken out. This exposes the clutch disc assembly as shown at Fig. 89, C. The clutch discs are carried by a member known as "the disc drum" shown in group of parts A, Fig. 89. A set screw passes through this member to key it securely to the crank shaft extension. When this set screw has been properly loosened the clutch disc drum may be removed which leaves the assembly as shown at Fig. 89, D. This assembly, consisting of the reverse drum, slow speed and brake drums and the triple gears, is held on the flywheel. This assembly may be easily withdrawn from the flywheel and the crank shaft extension which is known as the "transmission" shaft which leaves the group as shown at Fig. 89, E. In order to take this down, the driven gear must be removed from the brake drum shaft extension, which will permit the three drums to be pulled apart.

The point to examine carefully after the transmission has been disassembled is to note if the bushings in the triple gear assemblies or if the pins attached to the flywheel used for supporting them are worn. If there is considerable play between the bushings and the pins the bushings should be forced out of the triple gear assemblies and new ones inserted in their place. The triple gear pins should also be replaced with new ones. The bushings in the reverse drum and gear and in the interior of the slow speed drum and gear should be carefully measured to make sure that the slow speed drum is a good fit on the extending shaft of the brake drum and that the reverse drum is a proper fit on the extension of the slow speed drum. If these bushings are worn so that considerable looseness obtains, the old ones should be driven out and new ones forced in under an arbor press.

Before reassembling the brake drum assembly, care must be taken to fit the bushings so they will turn freely on the members by which they are supported. The relation of the parts to each other in the complete assembly can be readily understood by referring to Fig. 27, which is a sectional view of the transmission gearing. The surfaces of the brake, slow speed and reverse drums should not be cut or scored as might result if the brake lining has been improperly applied and iron rivets used instead of those supplied by the Ford Company. It is not only important to use the proper rivets but they should be properly countersunk.

The transmission clutch discs should be removed from the assembly, thoroughly cleaned and inspected to see if these are smooth or if the surfaces are badly scored. If the contacting faces are ridged, which would result if the operator continually slipped the clutch, the discs should be smoothed off if the ridges are not deep and new ones used if the surfaces are badly scored. While taking the transmission gear apart is not a difficult operation, some skill is needed to insure correct reassembling.

The first operation is to assemble the driven gear, reverse drum and gear, slow speed drum and gear and brake dram, all of which are shown at Fig. 89, A, to form the group shown at Fig. 89, E. Place the brake drums on a bench or table with the hub extending vertically, then place the slow speed drum over this hub with the gears uppermost. The reverse drum is then assembled over the hub of the slow speed drum with its gear member uppermost. Next, place two Woodruff keys utilized to driving the pinion marked "Driven Gear" shown at A, Fig. 89, in the key ways cut into the brake drum hub just above the slow speed gear. Put the driven gear in place with the teeth downward so they will come next to the slow speed gear. Take the three triple gears and mesh them with the driven gear so that the punch marks on the teeth correspond, the reverse gear, or smallest one of the three comprising the assembly, being downward. When the triple gears have been properly meshed they should be tied in place by passing a cord or wire around the outside of the three gear assemblies which is cut and removed when the group is in place. This group then has the appearance shown at Fig. 89, E.

The next step is to assemble group E on the flywheel. The flywheel is placed on the bench with its face downward so that the transmission shaft projects vertically. Group E is inverted so that the triple gear assembly will face the flywheel, then the group is pushed on the transmission shaft allowing it to settle in such a position that the triple gear supporting pins on the flywheel will pass through the bushings in the triple gear assembly. This will bring the brake drum on top as shown at Fig. 89, D. The next thing to do is to fit the clutch drum key in the transmission shaft and press the clutch disc carrier drum in place on the shaft, locking it into position with a setscrew provided for that purpose. The distance plate, which is a heavier disc than the clutch plates, is put on the clutch drum first, then a small disc which is followed by a large one, then placing another small disc and alternating large and small ones until the entire set of discs is in position, a large one or member having keyways in its outer periphery, being on top. Care should be taken never to have a small disc or one with keys cut in its inner periphery on top because it is liable to fall over the clutch drum when changing the speed from high to low, and as a result one would be unable to re-engage the high speed clutch.

The appearance with the clutch disc drum and the clutch discs in place is shown at Fig. 89, C. The next step is to put the clutch disc ring over the clutch drum, then put the clutch push ring over the clutch drum and on top of the disc ring with the three pins projecting upward as shown in group of parts B, Fig. 89. The remaining parts to be assembled are placed in the order to be followed in replacing them. Bolt the driving plate in position on the brake drum so the adjusting screws of the clutch fingers will bear against the clutch push ring pins. Before proceeding further the Ford Company advises the worker to test the transmission by moving the plates or drums with the hands. If the transmission is properly assembled the flywheel will revolve freely while holding any of the drums stationary or vice versa.

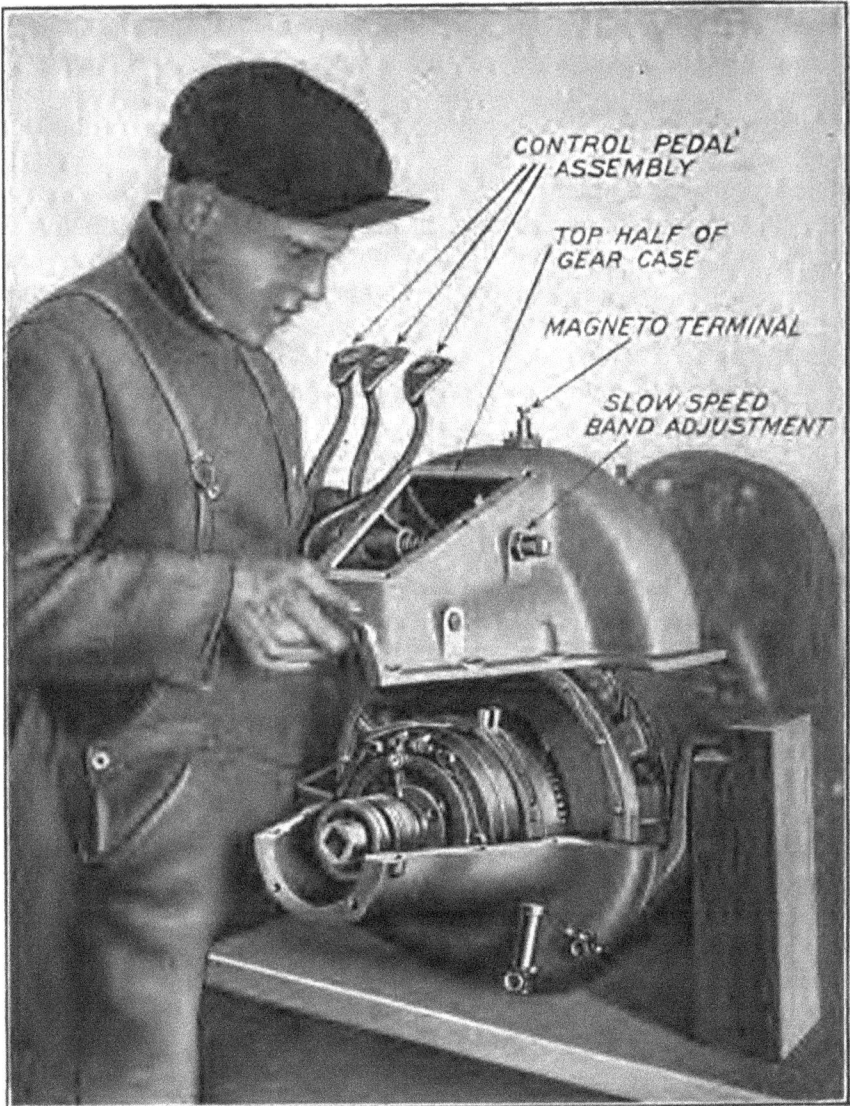

Fig. 90.—Showing How Transmission Cover is Removed to Permit Reaching Transmission Brake Bands.

The clutch parts may be assembled on the driving plate hub as follows: Slip the clutch shift on the hub so the small end rests on the ends of the clutch finger, next put on the clutch spring with the clutch support inside so the flange of that member will rest on the upper coil of the spring. Next place clutch spring thrust ring with notched end down and press into place, inserting the pin in the driving plate hub through the holes in the side of the spring support. The easiest method of compressing the spring sufficiently to insert this pin is to loosen the clutch finger tension by backing out the adjusting screw. When tightening up the clutch again the spring should be compressed to a length of 2 or 2 1/16 inches to insure against the clutch discs slipping. The precaution should be taken to see that the screws in the driving plate fingers are adjusted uniformly in order to obtain even compression of the clutch spring.

Relining Brake Bands. — The parts of the transmission gear that will wear soonest and which will need inspection and repair long before the bushings of the gearing have worn are the friction linings in the three transmission brake bands. The only way a new lining can be put on is to remove the brake band from the transmission assembly, which can be done without taking the power plant out of the frame. The first step is to take off the door on top of the transmission cover and to turn the reverse adjustment nut and the brake adjustment nut to the extreme end of the thread on the pedal shaft. This permits the brake bands to expand away from the drum. It is also important to slack off the slow speed adjusting screw. Next remove the bolts holding the transmission cover to the crank case and lift off the cover assembly as shown at Fig. 90.

Slip the band nearest the flywheel over the first of the triple gears, then turn the band around so that the opening is downward. The band can now be removed by lifting upward, as shown at Fig. 91. The operation is more easily accomplished if the three sets of gears are so placed that one set is just a little to the right of center at the top. Each band is removed in the same way. It is necessary to push each band forward onto the triple gear assemblies as it is only at this point that there is room enough in the crank case to allow the upstanding slotted ears on the transmission bands to be turned down. The bands are replaced by reversing the operation. After being placed in the upright position on drum, a cord is passed around the ears of the three bands so that when putting on the transmission cover no trouble will be experienced in having the pedal shafts rest in the notches made to receive them in the band ears. The clutch release ring must be placed in the rear groove of the clutch shift ring. When the cover is in place remove the cord that held the bands together while the cover was being installed.

The operation of relining a brake band is a relatively simple one, consisting only of pulling off the old brake lining and driving out the retaining rivets from the holes in the band. A new lining is placed inside the band and a piece of steel bar stock is placed in the vice as shown in Fig. 92, to form a backing against which the special rivets are driven by means of a steel drift or rivet set to clinch the lining securely in place. The linings furnished by the Ford makers are the only ones that should be used because serious ignition trouble may be occasioned by particles of wire or impregnating compounds dropping off of linings not made for this purpose getting around the magneto current collecting plunger.

Fig. 91.—Removing Transmission Brake Band for Replacing Friction Lining.

Rear Axle Troubles and Remedies. — If continual grinding noises are heard in the rear axle when the car is being operated on the high speed drive, this is a sign that the gears or the bearings in the rear axle are worn and calls for a thorough overhauling of that member. Wear in bearings can be tested by jacking up the rear end and lifting upon the wheels in the same way as advised for testing the emergency or rear hub brakes. If it is possible to move the wheel up and down it indicates that either the roller bearings or the axle is worn. Steps should be taken to correct this, because wear at this point imposes great strains on the roller bearings at either side of the differential.

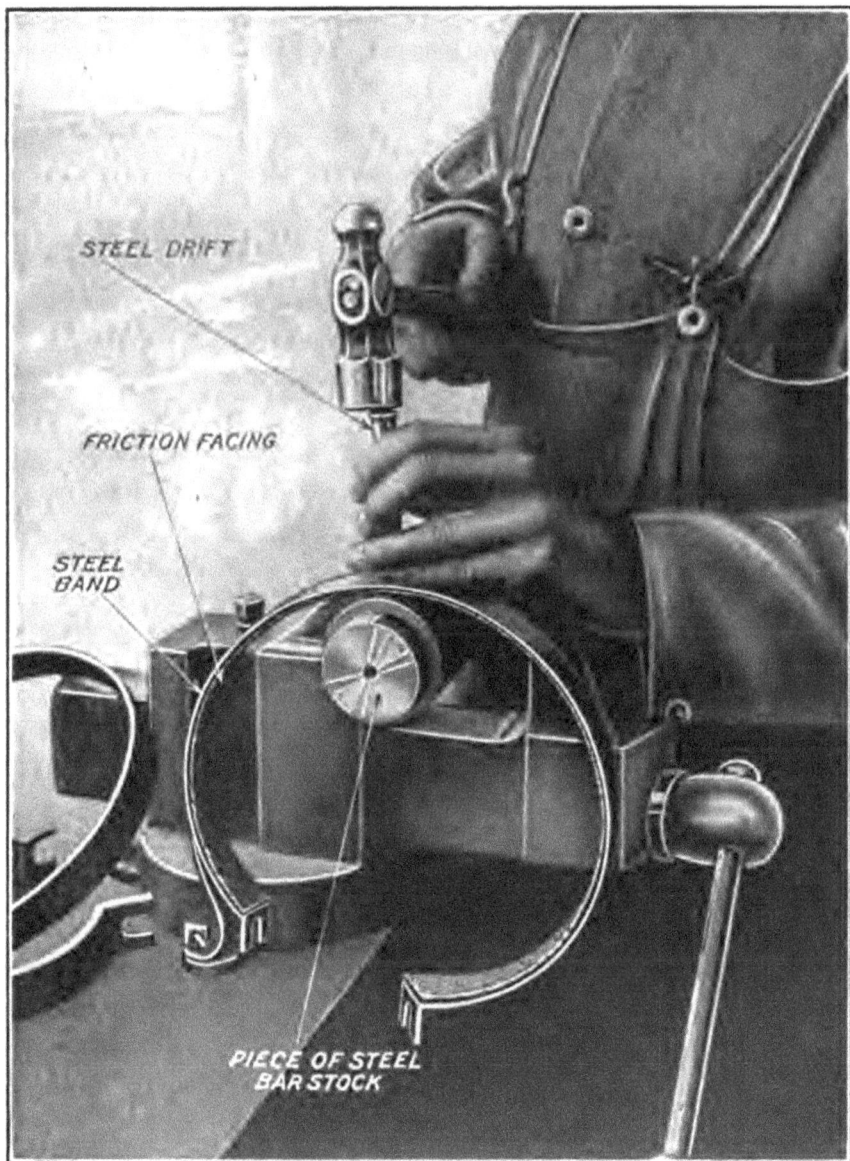

Fig. 92.—How Friction Lining is Riveted to Brake Bands.

After a car has been in use for a time and especially if the differential housing has not been kept properly lubricated, the babbitt thrust collars at each side of the differential and the bronze bushings in the differential case may wear sufficiently so that the bevel driving gears will not mesh properly. The best way of making axle repairs is to take this member apart after it has been removed from the chassis and carefully examine all worn parts. The first step in removing the rear axle is to jack up the car and remove the rear wheels as previously described for inspecting the hub brakes. Take out the four bolts connecting the universal joint ball retaining cap to the transmission case and cover. Disconnect the hub brake rods from the cross shaft to which the hand lever is fastened. Raise the frame so that the weight will be released which will permit of removing the spring shackles or the spring perches from the rear axle housing flanges if this method is preferred.

It is not necessary to remove the wheels while the axle is under the car as the spring may be disconnected from the spring perches by taking off the spring shackles. Some repairmen prefer to leave the wheels on as it enables them to roll the axle out from under the car. The first step in taking down the axle is to remove the retaining nuts at the end of the radius rods. Then to re- move the drive shaft tube by taking off the nuts on the six retaining studs holding the drive shaft assembly to the rear axle housing. Next remove the bolts which hold the two halves of the differential housing together. The wheels must be removed from the axle shaft to permit one to spread the housings apart as shown at Fig. 93, and obtain access to the differential gear. The two halves of the axle housing may be removed entirely, if desired, thus leaving the differential gear with the two wheel drive shafts extending from it, one at either side.

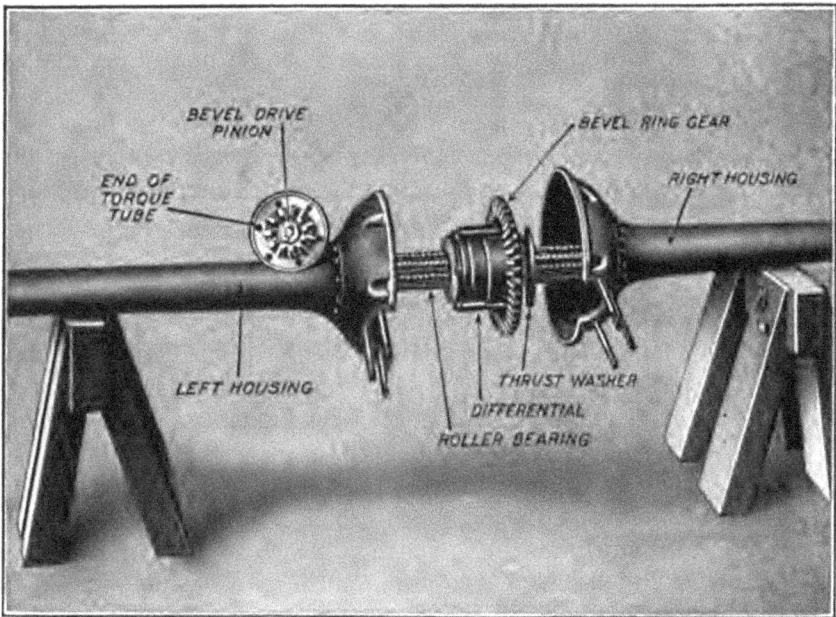

Fig. 93.—Rear Axle Partially Disassembled to Show Differential and Supporting Bearings.

The gear teeth should be carefully examined, especially those on the bevel drive pinion which are apt to wear more quickly than those of the bevel ring gear on account of their being a lesser number to transmit the power. If the

163

pinion teeth surfaces have flaked or if portions of the teeth are broken, a new pinion should be installed in place of the defective one. Before installing a new pinion care should be taken to see that there is no lost motion in either the ball or roller bearings, back of the pinion. Any play in these members must be removed by fitting new bearings because these are the most important bearings in the entire axle mechanism. Any play in these bearings may result in loss of alignment between the driving and driven gears.

The pinion may be easily removed from the end of the drive shaft by taking out the locking pin passing through the castellated nut and removing the nut, then drawing the pinion off the driveshaft with a suitable gear puller. The babbitt thrust washers and the steel plates on each side of them should be examined to see that these are not worn and the axles should be grasped and moved up and down to see if there is any play between them and the differential casing. If any looseness is present at this point it is because the bronze bushings in the differential casing that bear on the hubs of the compensating gears have worn. The differential casing is taken apart by unloosening the through bolts which frees the two halves of the casing. This exposes the compensating gears, and if these have been damaged as might result if a piece of broken pinion tooth was to get into the differential mechanism, it is necessary to remove the compensating gears from the axle shaft.

On examination it will be seen that the gears are keyed on and are held in position by a ring which is in two halves and which fits in a groove in a shaft. To remove the gears they must be forced down on the shaft away from the end to which they are secured in order to expose the rings which are removed from the grooves to permit the gears to be forced off the end of the shaft. If the roller bearings have worn and the shaft is not cut or worn in at the bearing point a new roll and cage assembly may be installed. Sometimes the steel lining or shell placed in the axle housing will wear. This may be easily removed and a new one inserted. If the axle shaft is reduced in diameter at the bearing point the only means of restoration possible is to replace the entire axle shaft. Leakage of oil or grease out of the axle is usually because the two halves are not securely bolted together if the leak is around the differential housing or because the felt retaining washers at the wheel end of the axles have depreciated. This may be suspected if the brakes have become fouled with oil.

Miscellaneous Chassis Parts. — There are a number of minor points about the chassis which demand inspection and in some cases adjustment when a car is overhauled. The most important of these is the front axle and steering connections. The front 'wheels should be inspected to . make sure that they are in proper adjustment and that they run smoothly without appreciable side play. The bearings should be thoroughly cleaned and looked over to see that there are no broken balls and that the cones and ball races have not become pitted or roughened on the ball track. If there is much play in the steering rods or in the bearings supporting the steering spindle bolt the only remedy is to drive out the worn bushings and replace with new ones, which will be an inexpensive operation.

If the front wheel is in proper adjustment it should spin easily and come to rest with the tire valve at the bottom. If the ball bearings in the front wheels wear out quickly it is usually due to water getting into the bearings, the use of lubricant containing acid or improper bearing adjustment. One can be sure that no water will get into the bearings if the felt washer on the inside of the wheel hub is in proper condition and if the front hub interior is filled with the proper grade of mineral grease. The rear wheels should be inspected after the car has been driven for a time to make sure that these fit the taper ends of the axle shafts tightly.

If there is considerable rattling and knocking at the front end of the car and the trouble is not due to loose engine bearings it is often caused by the ball joint at the end of the front axle radius member being loose. This may be tightened up by removing the cap and grinding off some of the cap face in order to have it set more tightly against the ball and to force that member into more intimate contact with the hemispherical seat on the front side of the flywheel compartment. Special spring adjusted caps are now furnished by the Ford Company. Rattling is also caused by loose steering connections and by loose mud guards.

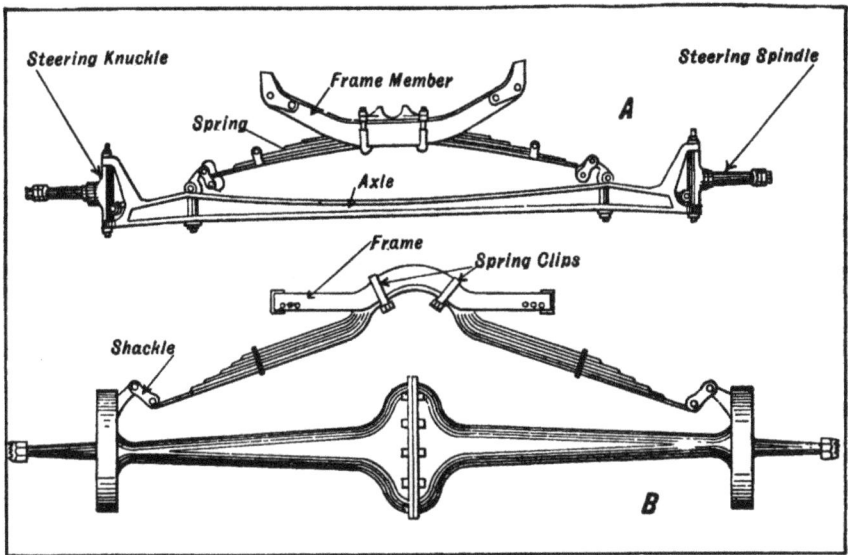

Fig. 94.—Outlining Method of Ford Front and Rear Spring Retention.

Squeaking noises result when the springs become rusty and no lubricant is present between the leaves. While the car is being overhauled it is a good plan to remove the springs from the chassis and to take these apart, cleaning off the surfaces with emery cloth and smearing them with a mixture of graphite and oil before reassembling. It is a very simple matter to remove the springs from the Ford frame as these are held by readily detachable spring clips, shown at Fig. 93. Both front and rear springs may be removed by taking off two spring clips and two spring shackles. It is important to jack up the frame so the weight will be taken off the springs before any attempt is made to remove these from the chassis.

After a car has been in service two or three years, excessive play in the steering gear may result from depreciation of the teeth of the small planetary pinions and internal gear mounted under the steering wheel spider. These must be replaced with new ones when worn. The steering wheel is removed from the steering column by unloosening the nut on top of the wheel spider and pulling off the spider hub from the shaft to which it is fastened. The interior of the steering gear may be easily inspected after the steering wheel is removed by loosening a set screw and unscrewing the brass cap that is a cover for the reduction gear case. If a wheel puller is not available for taking the steering wheel off of the steering post it may be driven off of the shaft with a block of wood and a hammer.

In addition to the main points mentioned there are a number of minor bearing points such as the bushings in the spring eyes, the shackle bolts and the various rod ends and pins at the joints of the control rods which will wear and produce their quota of noise. The bolts employed for holding the body to the chassis sometimes loosen, this resulting in quite a severe pounding noise when the car is operated over anything but the smoothest highway.

www.ingramcontent.com/pod-product-compliance
Lightning Source LLC
Chambersburg PA
CBHW030527100426
42813CB00001B/171